REDEFINING INFORMATION WARFARE BOUNDARIES FOR AN ARMY IN A WIRELESS WORLD

T0127970

Isaac R. Porche III, Christopher Paul, Michael York,
Chad C. Serena, Jerry M. Sollinger, Elliot Axelband,
Endy Y. Min, Bruce J. Held

Prepared for the United States Army
Approved for public release; distribution unlimited

ARROYO CENTER

The research described in this report was sponsored by the United States Army under Contract No. W74V8H-06-C-0001. The findings and views expressed in this report are those of the authors and do not necessarily reflect the views of the Army or the U.S. Department of Defense.

Library of Congress Cataloging-in-Publication Data is available for this publication.

ISBN 978-0-8330-5912-3

The RAND Corporation is a nonprofit institution that helps improve policy and decisionmaking through research and analysis. RAND's publications do not necessarily reflect the opinions of its research clients and sponsors.

RAND® is a registered trademark.

Published 2013 by the RAND Corporation
1776 Main Street, P.O. Box 2138, Santa Monica, CA 90407-2138
1200 South Hayes Street, Arlington, VA 22202-5050
4570 Fifth Avenue, Suite 600, Pittsburgh, PA 15213-2665
RAND URL: http://www.rand.org/
To order RAND documents or to obtain additional information, contact
Distribution Services: Telephone: (310) 451-7002;
Fax: (310) 451-6915; Email: order@rand.org

Preface

As the Army observed in the 2010 cyberspace operations concept capability plan, society's dependence on the wireless and wired mediums is converging. Computer and telecommunication networks are becoming one and the same. And the transmission of digitized packets on Internet-protocol and space-based networks is rapidly supplanting the use of old technology (e.g., dedicated analog channels) when it comes to information sharing and media broadcasting.

This monograph identifies the implications of these trends and reconsiders the resulting boundaries of Army cyber operations, at least from a practical standpoint. It focuses on the general and overlapping areas of network operations, information operations, and the more focused areas of electronic warfare, signals intelligence, electromagnetic spectrum operations, public affairs, and military information support operations (formerly psychological operations). Most importantly, it compares the emerging doctrine of cyber operations to all of the aforementioned areas. The intent is to make clear the prevailing boundaries between the areas of interest and the expected progression of these boundaries in the near future. It constructs some new definitions that encapsulate these areas, such as information warfare. This is important because the Army is now studying ways to best apply its cyber power and reconsider doctrinally defined areas that are integral to cyberspace.

This monograph asserts that the relevant realms that contain the functional areas pertaining to information warfare are just two: the psychological and the technical. The psychological is focused on message content, and the target is people. The technical realm is focused on the means to deliver (or prevent delivery of) content, and the targets are machines. This monograph considers how the technical realm and the psychological realm can best be organized and perhaps consolidated.

This study and monograph were not specifically requested by the Army; rather, this monograph summarizes the results of a short study conducted in response to a question about the future of information operations asked by Army senior leadership. RAND Arroyo Center sought an answer to this question as a "Quick Response" study. Quick Response studies are designed to support near-term decisions to be made by Army officials or to provide analyses to the Army leadership to inform U.S. Department of Defense, administration, or congressional decisions and actions. A brief was provided to Army senior leaders within two months of initiation of this project; this monograph summarizes and reports the analytic effort that went into that briefing. The findings and views expressed here are those of the authors and do not necessarily reflect the views of the Army or the U.S. Department of Defense.

This research was conducted within the Arroyo Center's Force Development and Technology Program. RAND Arroyo Center, part of the RAND Corporation, is a federally funded research and development center sponsored by the U.S. Army. Questions and comments about this research are welcome and should be directed to the program director, Christopher Pernin (Christopher_Pernin@rand.org), the project leader, Isaac Porche (Isaac_Porche@rand.org), or Christopher Paul (cpaul@rand.org).

For more information on RAND Arroyo Center, contact the Director of Operations (telephone 310-393-0411, extension 6419; FAX 310-451-6952; email Marcy_Agmon@rand.org), or visit Arroyo's website at http://www.rand.org/ard. The Project Unique Identification Code (PUIC) for the project that produced this document is RAND10473.

Contents

Figures

Tables

Summary

Information warfare is not currently defined in U.S. Department of Defense (DoD) or U.S. Army doctrine, but it is a term found in past doctrine.[1] What is in today's DoD lexicon is the term *information environment*, the "aggregate of individuals, organizations, and systems that collect, process, disseminate, or act on information" (U.S. Joint Chiefs of Staff, 2010b). Joint doctrine (e.g., JP 3-13.1) makes clear that "there is an electromagnetic spectrum portion of the information environment" (U.S. Joint Chiefs of Staff, 2007, p. vii).[2] Thus, wired and wireless technology fit in this landscape.

As a term, *information warfare*, or IW, remains in use worldwide, in the militaries of other countries as well as in some of the U.S. military services. The Navy now has an IW officer position, which it advertises as involving "attacking, defending and exploiting networks to capitalize on vulnerabilities in the information environment" (U.S. Navy, undated). Career paths for these officers are described in Appendix F. We define IW as follows:

> *Information warfare is conflict or struggle between two or more groups in the information environment.*[3]

---

[1] There is no entry in Joint Publication 1-02 (U.S. Joint Chiefs of Staff, 2010b). *Past doctrine* here refers to the mid-1990s. See AFDD 5, 1996, and CJCSI 3201.01, 1996.

[2] Joint doctrine says that a portion of the information environment includes the electromagnetic environment (EME). See U.S. Joint Chiefs of Staff, 2007.

[3] Dan Kuehl of the National Defense University defines IW as "military offensive and defensive actions to control/exploit the environment" (various briefings); U.S. Joint Chiefs

Social networks, as part of the information environment, are also a part of such conflicts or struggles. As noted by LTG Michael Vane, "Army forces operate in and among human populations, facing hybrid threats that are innovative, networked, and technologically-savvy" (TRADOC, 2010a, p. i).[4] Internet-assisted social networking is now a part of the operational environment, as events in Egypt, Moldova, Iran, and even Pittsburgh have made clear.[5] Social networks are a growing and increasingly relevant element of the information environment.

Cyberspace is the technical foundation on which the world is increasingly relying to exchange information (and to facilitate social networking, extend influence from afar, and so on). As a collection of mediums, it is rapidly consuming the information environment's

---

of Staff (1995) notes that "IW focuses on affecting an adversary's information environment while defending our own." CJCSI 3210.01 (1996) defined information warfare as follows: "Actions taken to achieve information superiority by affecting adversary information, information-based processes, information systems, and computer based networks while defending one's own information, information-based processes, information systems and computer-based networks."

[4]   On September 16, 2010, Deputy Secretary of Defense William J. Lynn III signed Directive-Type Memorandum 09-026 establishing Internet-based capabilities as an integral part of DoD operations. Falling under the realm of Internet-based capabilities is social media.

[5]   Cell phones and text messaging are believed to have played a crucial role in fostering the so-called Orange Revolution in the Ukraine. Twitter is credited with making these protests widespread and successful (e.g., flash mobs). Ultimately, the protests forced a recount of the general election. See Morozov (2009), Goldstein (2007), and Stack (2009).

During Iran's so-called Twitter revolution, it was reported that well-developed Twitter lists showed a constant stream of situational updates and links to photos and videos, all of which painted a portrait of the developing turmoil. According to news reports, when the Iranian regime started taking down these sources, the so-called e-dissidents shifted to email. (See "Iran's Twitter Revolution," 2009.)

During a recent G20 meeting, protesters in Pittsburgh leveraged Twitter. For example, Elliot Madison, an activist in New York City, used Twitter to disseminate information about Pittsburgh police activities and movements during the protests. Reportedly, police raided Madison's hotel room, and, one week later, his home was raided by FBI agents. Police reports claim that Madison and a co-defendant used computers and a radio scanner to track police movements and then passed that information to protesters using cell phones and Twitter. Madison is reportedly being charged with hindering apprehension or prosecution, criminal use of a communication facility, and possession of instruments of crime (Democracy Now! 2009; Electronic Frontier Foundation, 2009; Goodman, 2009).

landscape. Therefore, we conclude that controlling cyberspace (and the intersecting electromagnetic spectrum) could eventually be tantamount to controlling the information environment. The Army must prepare for that possibility.

The Problem with Current Doctrine

Preparation for IW will start with revision of the 2003 Army Field Manual (FM) 3-13, *Information Operations* (IO), which is widely considered antiquated and insufficient for the future. Harkening back to the birth of the information operations concept out of command and control warfare in the late 1990s, this doctrine aggregates the areas of electronic warfare (EW),[6] computer network operations (CNO), psychological operations (PSYOP),[7] military deception (MILDEC), and operations security (OPSEC) as core capabilities, despite the fact that some of these concepts are quite dissimilar. This is shown in Figure S.1.

One conflict that has emerged stems from overlapping doctrine. For example, CNO, historically covered in FM 3-13, is the main component of cyber operations. According to the latest Army operating concept, "Cyberspace operations include computer network operations" (TRADOC, 2010b). Similarly, EW has its own doctrine (FM 3-36) and a growing force structure. Thus, we can say that the growth in size and importance of EW, CNO, and cyber operations as a whole render them too large and fast-moving to fit within this IO doctrine.

The confusion associated with IO as a term—in the Army and at the joint level—stems from many sources: genuine ambiguity in the lexicon, both willful and unintentional misuse of the term, and both genuine misunderstanding and genuine disagreement about what such operations are and how they ought to be defined.

---

[6] Certain functions in EW can be considered military deception. This includes the use of expendibles (e.g., flares) by vehicles (Hura, 2010). This should be (and likely is already) included in EW doctrine and/or corresponding tactics, techniques, and procedures.

[7] Now referred to as military information support operations (MISO). See Chapter Two.

Figure S.1
IO Doctrine

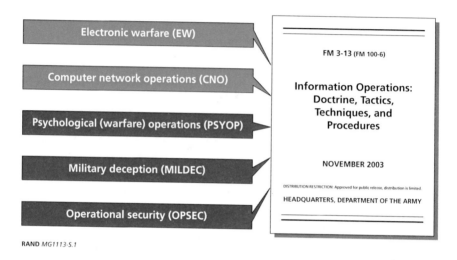

Electronic warfare (EW)

Computer network operations (CNO)

Psychological (warfare) operations (PSYOP)

Military deception (MILDEC)

Operational security (OPSEC)

FM 3-13 (FM 100-6)

Information Operations:
Doctrine, Tactics,
Techniques, and
Procedures

NOVEMBER 2003

DISTRIBUTION RESTRICTION: Approved for public release, distribution is limited.

HEADQUARTERS, DEPARTMENT OF THE ARMY

RAND *MG1113-S.1*

As reflected in Figure S.1, at the time of this research, joint doctrine defined IO as follows:

> The integrated employment of the core capabilities of electronic warfare, computer network operations, psychological operations, military deception, and operations security, in concert with specified supporting and related capabilities, to influence, disrupt, corrupt or usurp adversarial human and automated decision making while protecting our own. (U.S. Joint Chiefs of Staff, 2007, p. G-10)

This definition does little to clear up the confusion, both because of ambiguities in the definition itself—because soldiers imagine it to (or want it to) mean something else—and because IO, as actually practiced, deviates from that definition.[8]

---

[8] This alone demands new doctrinal writings. As Maj Gen I. B. Holley (1983) said, "What is doctrine? Simply this: doctrine is officially approved prescriptions of the best way to do a job. Doctrine is, or should be, the product of experience. Doctrine is what experience has shown usually works best."

In January, 2011, Secretary of Defense Robert Gates issues a memorandum outlining a revised definition of IO, with a greater focus on integration. He stated that the definition in effect when this research was conducted placed "too much emphasis on core capabilities" and supported the "notion that the core capabilities must be overseen by one entity. Joint doctrine now defines IO as

> the integrated employment, during military operations, of information-related capabilities in concert with other lines of operation to influence, disrupt, corrupt, or usurp the decision-making of adversaries while protecting our own. (Gates, 2011)

There are genuine disputes regarding both the terminology and the concepts of IO, so resolution cannot be had with simple clarification. There are decisions to be made.

Information Operations as a Moving Target

Further complicating the situation is that the need for change in IO is recognized, and progress is under way as of this writing in both the joint community and the Army toward improving definitions, revising doctrine, clarifying concepts, and adjusting organizations. The 2011 Gates memo is an example of the progress made between the time that this monograph was first drafted and the time of its publication. The authors have endeavored to stay abreast of such movement, but other changes and advances have taken or are taking place at the time of publication. Undoubtedly, some important decisions will have been made, other important progress will have occurred, and some of the recommendations presented here will have been overtaken by events.

However, challenges will remain. Debate within and surrounding the information operations community runs hot and fierce, and progress is often delayed by disagreements (such as the 2009 attempt to revise FM 3-13, described in Appendix B). The information environment continues to evolve, adding new challenges. If we believed that all the issues facing information operations would be resolved by the time this monograph was released, we would not have published it. As the doctrine and practice of IO continue to evolve, this monograph will remain useful when further changes are considered or when past changes are revisited, reviewed, and debated again.

Resolving the Problems by Redefining Doctrinal Terms

More clarity can be provided by separating the functional areas currently defined in the IO definition into two realms: the more technical functional areas and the other functional areas associated with PSYOP/MISO.

Information content (e.g., the message) is key for the psychological part; the means to deliver content (or prevent delivery) is key for the technical part. Ultimately, it makes sense that most of what falls into the psychological realm (shown in Table S.1) be redefined as inform and influence operations (IIO) and that most of what falls into the technical realm be considered information technical operations (ITO).

Essentially, we suggest that the doctrine be split to reflect how the expertise has been divided today, as illustrated in Figure S.2. The table does not account for the integrating function; integration of these areas belongs with the commander. Revisions to the mission command doctrine should reflect this.

Network operations fall clearly within the technical realm. While we include them in the table for completeness, we do not foresee any practical benefit in merging network doctrine and personnel with the other areas. This is because network operations are large, long-standing efforts in the Army that should remain focused (Porche et al., 2010). However, this issue requires more study and was outside the scope of our research.

We offer the following definition of IIO. It integrates features of three different visions of what IO could be (these different visions are described in Chapter Three): an integrating function, an influence capability, and an advisory capability.

> *Inform and influence operations are efforts to inform, influence, or persuade selected audiences through actions, utterances, signals, or messages.*

Table S.1
Information Warfare: Realms of the Possible

Category	Psychological Realm	Technical Realm
Functional areas, subareas, defined in existing doctrine	MISO, public affairs (PA), aspects of MILDEC	Electronic attack (EA), electronic protect (EP), electronic support (ES), computer network attack (CNA), computer network exploit (CNE), signals intelligence, electromagnetic spectrum operations (EMSO), information assurance, operating and maintaining networks (network operations), aspects of MILDEC, aspects of OPSEC
Target	People	Machines
Alternate name	Inform and influence operations (IIO)	Information technical operations (ITO) or cyber-electronic operations or cyber-electromagnetic operations

Figure S.2
The Dividing Line That Should Be Sharpened: Technical Operations Versus Inform and Influence Operations

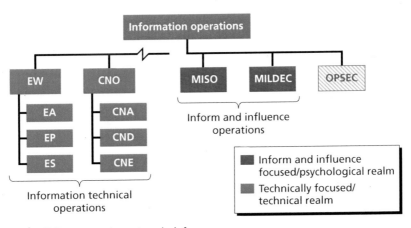

NOTE: CND = computer network defense.

RAND *MG1113-S.2*

We offer the following definition of ITO. We do not mean to imply that all or most of the areas are covered under signals intelligence. It is a combination of electronic warfare, computer network operations, and other functions.

> *Information technical operations are efforts to protect and/or coordinate U.S. and allied technical means and mediums (e.g., the EMS) that facilitate command-and-control and, perhaps, certain intelligence activities and to deny the means and mediums used by adversaries.*

This definition and the associated vision have several notable characteristics. First, this definition separates the "apples" of information content from the "apple carts" of information systems (e.g., information technology and electronics) and retains the term *information operations* to refer to the former exclusively (see Paul, 2008). Under this vision, IIO include only efforts to inform, influence, or persuade.

Advantages of Revising Doctrinal Definitions

These separated definitions make clear the distinctions between the functional area groupings (i.e., the psychological and technical realms) shown in Table S.1. More distinction helps lessen the confusion that exists today regarding who executes the missions shown in Figure S.1. The personnel in these areas might be more focused and better able to develop concentrated expertise. Finally, separating these areas could translate into more opportunity to consolidate within them.

Consolidation in the Technical Realm

Consolidation in the technical realm is possible and advisable. The boundary between CNO over wireless networks and EW is blurring. At a minimum, the impact of the convergence trend is that EW (electronic attack [EA], electronic protect [EP], and electronic support [ES]) and CNO (computer network attack [CNA], computer network defense [CND], and computer network exploitation [CNE]) are becoming increasingly comingled.

On the materiel side, the convergence of wired and wireless mediums suggests that there might be circumstances in which the functional requirements of these currently separate areas can be met by the same device that combines technologies to yield the best system solution. Advanced electronic steerable array (AESA) radars might fall into the EW and the CNO areas because they can sense and transmit in both analog and digital formats.

As a result, we conclude that EW and CNO could—and perhaps should—share the same people, process, and technologies to carry out these operations to avoid duplication of effort or working at cross-purposes. We understand that the Army has already begun to make some moves toward aggregation in this area.

Proposals already exist to merge existing EW and emerging cyber operations doctrine, and they appear to be advantageous; progress is being made in this direction. However, there are cautions. Today, the authorities required to conduct offensive EW are more clearly understood and more permissive than the authorities that exist for offensive CNO. In addition, the clearance levels required for offensive EW differ from those required for offensive CNO.[9] Doctrine for EMSO and spectrum managers themselves (e.g., personnel with the 25E military occupational specialty [MOS])[10] should be part of this consolida-

---

[9] For example, generally, electronic attack operations are planned at the secret level, and authority to plan and execute operations resides with tactical- and operational-level commanders. On the other hand, CNA operations are often conducted at higher security levels (Hura, 2010).

[10] The signal corps' MOS 25E enlisted specialty for spectrum management was created a number of years ago. Prior to the creation of this specialty, noncommissioned officer spectrum managers were tracked only with a skill identifier attached to a preexisting MOS. The skill identifier for enlisted personnel (for spectrum managers) was not found to be satisfactory because these spectrum managers were often retasked outside of the spectrum specialty. There is a skill identifier for commissioned officers, but it is dormant.

In the case of EW, the Army recently created a new career management field that provides a new MOS for officers, warrant officers, and enlisted personnel. Hundreds of billets (greater than 3,000 personnel) have been created, although not all have been filled. The specific career management field identifiers for EW are to be FA29 for officers, MOS 290A for warrant officers, and MOS 29E for enlisted personnel.

tion. We illustrate this proposal in Figure S.3. Also, technical aspects of OPSEC and MILDEC fall here.

The Army eventually needs to either create a new "cyber-electronic" or "cyber-electromagnetic" career management field or transform an existing one (e.g., CMF 29) to provide dedicated support to all or most of the technical realm of IW. This would serve as a first step toward a new branch for cyber-electromagnetic warriors for the far future who can be utilized to cover the areas discussed here. This group includes EW and spectrum managers.

Consolidation in the Psychological Realm

A similar argument can be made for the psychological realm, where there is just as much opportunity for consolidation. Specifically, PA and MISO (formerly PSYOP) have ample reason to become better integrated.

Currently, there is a "firewall" between PA and MISO. The concern that has kept PA and MISO separate is the commitment to use (or not use) truthful information. However, the lack of PA-MISO coordination has resulted in repeated instances of "information fratricide," in which the separate capabilities provide conflicting information. The fear is that MISO could contain less-than-truthful information and thus jeopardize the credibility of PA efforts. However, almost all conventional MISO use truthful information (and sometimes the only dif-

Figure S.3
Potential Consolidation in the
Technical Realm

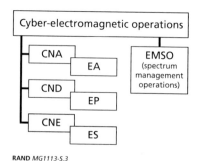

ference is the audience). A commitment to the truth is a reasonable approach. A bridge between the two seems possible with the approach suggested here.[11] Thus, beneficial integration and perhaps even consolidation (organizationally and/or with respect to personnel) is conceivable as envisioned in Figure S.4. At minimum, the firewall between inform and influence should be removed and placed at the bounds of truth and good intention. All communications seek to influence, and that is OK. Where the line should be drawn is between truthful efforts at virtuous persuasion (wholly acceptable) and deceptive manipulation.

Courses of Action

We see three courses of action for revising Army doctrine, addressing needs of FM 3-13 and the use of information operations officers.

1. The first course of action is to maintain the status quo (nearly). Maintain a broad definition of IO that involves integration of the five key functional areas listed in Figure S.1. This course of action includes continued reliance on FM 3-13 nearly as is;

Figure S.4
Potential Consolidation in the Psychological Realm

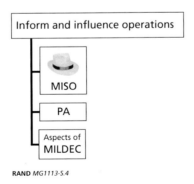

RAND *MG1113-S.4*

---

[11] NATO tried unsuccessfully to merge IO and PA but without making a clear commitment to use the truth only.

it would perhaps need an update to accommodate the expected revision to the joint definition of IO (when it becomes official).

2. Course of action two is to develop new doctrine that divides current doctrine (FM 3-13, as shown in Figure S.1) into IIO and ITO, as suggested in Figures S.2, S.3, and S.4. Integration functions would explicitly become the task of commanders, and this role and task will have documented in the corresponding doctrine (mission command). As such, FM 3-13 would be obviated, as would the role of IO officers as integrators. Doctrinally, ITO would fall under cyber-electronic operations or cyber-electromagnetic operations, which could be addressed in a revised FM 3-36 (currently titled *Electronic Warfare*).

3. Course of action three is to limit the scope of IO and IO officers to IIO as we define it here. This would essentially involve redefining IO to include only the functional areas we list in the psychological realm and the integration role of IO officers to be one of integrating MISO, PA, MILDEC, and similar functional areas.[12] To be clear, this would involve redefining IO *as efforts to inform, influence, or persuade selected audiences through actions, utterances, signals, or messages.* Doctrinally, ITO would fall under cyber-electronic operations or cyber-electromagnetic operations, which could be addressed in a revised FM 3-36 (currently titled *Electronic Warfare*).

Recommendations

Based on our review of the literature and analysis of overlapping tasks in some of these functional areas, we recommend course of action two. This is illustrated in Figure S.5.

Not addressed explicitly by the listed courses of action but certainly important to the discussion is the relationship of OPSEC as a

---

[12] For example, revise FM 3-13 to cover the need to integrate only MISO, PA, MILDEC, and other capabilities contributing to informing and influencing. Retain FA30s to integrate inform and influence. Certain aspects of MILDEC fall under EW (e.g., use of expendables and flares).

capability area to the new structure. We believe that OPSEC is every-one's responsibility. Aspects of it certainly fall under technical opera-tions, but it could also be covered in the mission command doctrine that is currently under revision.

Figure S.5
Recommended Course of Action: Redefine Information Warfare Operations

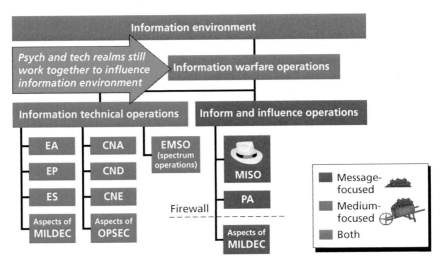

RAND *MG1113-S.5*

Acknowledgments

We benefited from extended conversations with personnel at the U.S. Army Communications-Electronics Research, Development and Engineering Center (CERDEC) Intelligence and Information Warfare Directorate (I2WD), including Giorgio Bertoli, chief of the Offensive Information Operations Branch; Paul Zablocky, chief engineer; Shawn Mathews; and Kevin Boyle. In addition, we consulted Seyhun Byrne, chief engineer of CERDEC Technology Initiatives.

We gained insight from a number of subject-matter experts on all aspects of IO: Professor Dennis Murphy, Austin Branch, COL Carmine Cicalese, LTC Russell Webb, LTC Amy McGrath, CDR (ret.) Michael Herrera, LTC Scott Riggs, and LTC Lawrence Chinnery. Our reviewers, Jeff Rothenberg, Lt Gen (ret.) Bob Elder, and RAND colleague Myron Hura gave graciously of their expertise and helped us improve the presentation of our research and findings.

We also had engaging discussions with LTC Karen Hillemheber and COL Michael Dominique at TRADOC, Ft. Leavenworth. Finally, we benefited from numerous conversations on this topic with COL R. Wayne Dudding, who generously shared his thoughts and experience.

Abbreviations

AESA	advanced electronic steerable array
C2W	command and control warfare
C3IEW	command, control, communication, intelligence, and electronic warfare
CBA	capability-based assessment
CERDEC	U.S. Army Communications-Electronics Research, Development and Engineering Center
CNA	computer network attack
CND	computer network defense
CNE	computer network exploit
CNO	computer network operations
COP	common operating picture
DoD	U.S. Department of Defense
DOTMLPF	doctrine, organization, training, materiel, leadership and education, personnel, and facilities
EA	electronic attack
EME	electromagnetic environment
EMS	electromagnetic spectrum

EMSO	electromagnetic spectrum operations
EP	electronic protect
ES	electronic support
EW	electronic warfare
EWI	electronic warfare integration
FM	Army Field Manual
I2WD	U.S. Army Communications-Electronics Research, Development and Engineering Center, Intelligence and Information Warfare Directorate
IA	information assurance
IEW	intelligence and electronic warfare
IIO	inform and influence operations
IO	information operations
ITO	information technical operations
IW	information warfare
JP	Joint Publication
LDO	limited-duty officer
MILDEC	military deception
MISO	military information support operations
MOS	military occupational specialty
OEF	Operation Enduring Freedom
OIF	Operation Iraqi Freedom
OODA	observe, orient, decide, act
OPSEC	operations security
PA	public affairs

PAO	public affairs officer
PSYOP	psychological operations
RCIED	radio-controlled improvised explosive device
RF	radio frequency
SBCT	Stryker brigade combat team
SIGINT	signals intelligence
TRADOC	U.S. Army Training and Doctrine Command

CHAPTER ONE

Introduction

Background

Since the creation of the Internet's predecessor, the ARPANet, the constant characteristic of the information environment has been one of kaleidoscopic change. A notable change in recent years has been the merging of the wired and wireless worlds as wireless technology becomes increasingly widespread and capable.

The rapid pace of change makes it difficult for even nimble corporations to keep up, and the challenge for the U.S. military is even greater. Acquiring materiel rapidly is difficult, given governmental controls and processes, and it is difficult to make rapid changes in the personnel structure. Thus, keeping up with major changes, such as the merging of the wired and wireless worlds, poses formidable challenges to the U.S. military.

Complicating the U.S. military's ability to accommodate change in the information environment is the fact that certain facets of that environment are not well understood. As a result, the organizations built to carry out military operations in the information environment are not ideal. Given that, as of this writing, the Army is revising its information operations (IO) doctrine, now is a perfect time to revisit how the Army has organized for such operations.

The Army's Role in Cyberspace

Cyberspace now pervades and has joined the traditional domains of conflict, including land, sea, air, and space (see Figure 1.1).

The U.S. Army plays an important and ever-changing role in cyberspace. Specifically, it operates and defends its own networks (network operations), allowing it to retain its freedom of action in the cyberspace domain. To respond to attacks and exploit opportunities, the Army's role must expand to include finding and targeting adversary networks that could affect U.S. military operations (network warfare). The vastness of the cyberspace domain cannot be overstated, and more resources are needed for the Army to operate there effectively. It needs more units with trained cyberwarriors.[1] In addition, Army organizations with cyber missions need the capabilities to operate coherently with each other and with joint and other agency partners and need to share a clear understanding of the rules of engagement.

Figure 1.1
Cyberspace as a Domain

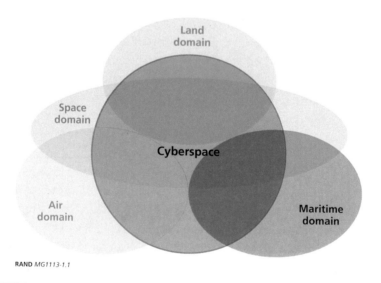

RAND MG1113-1.1

---

[1]   We use the term *cyberwarrior* to refer to personnel who carry out operations in cyberspace. Specifically, we define a cyberwarrior as someone who performs tasks that are vital for network operation, network defense, network attacks, or network exploitation.

The U.S. Department of Defense (DoD) and the military services have increasingly relied on networks to carry out and facilitate military operations. This is due, in part, to two trends. First, information operations are playing an increasingly important role in current operations and the migration into what we call *cyberspace*. The second trend is the increasingly prominent role that networks play on today's battlefield and their critical place in the Army's strategy and doctrine. These trends and the dependencies that they create have attracted the notice of adversaries, who view the expansion of cyberspace and the United States' dependence on it as creating military opportunities that they would not have in the traditional domains and are beefing up their own cyberspace capabilities to operate offensively and defensively in this emerging arena.

In response, the Army has been devoting resources to its own cyber capabilities, and, if future acquisitions go as planned, that investment will grow. Currently, however, the Army's cyber capabilities fall short of its needs. This paper addresses some of those shortcomings and provides some suggestions about what the Army can do to narrow the gap between its needs and its capabilities.[2]

What Is Cyberspace?

Environments

Traditionally, three military environments have formed the sphere of military conflict: atmospheric, terrestrial, and maritime. More recently, space was added as a military environment (see Figure 1.2). These environments are defined by their physical characteristics and include interfaces with the other environments, as suggested in the figure.[3]

---

[2] This discussion is informed by a yearlong study of the Army's role in cyberspace and prior studies requested by the Army's Chief Information Officer.

[3] DoD defines the terrestrial environment as "the Earth's land area, including its manmade and natural surface and sub-surface features, and its interfaces and interactions with the atmosphere and the oceans" (U.S. Joint Chiefs of Staff, 2010b).

Figure 1.2
Traditional Military
Environments

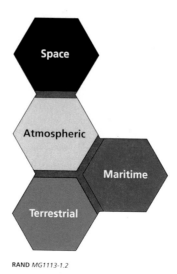

RAND *MG1113-1.2*

Domains

No official definition exists in the military for *domains*, but their usage in military publications suggests that they are broader than the environments and are areas of both operation and responsibility. As a result, military domains overlap (as illustrated by the ovals in Figure 1.3) and align with a separate service being responsible for each domain.[4]

Information Environment

With the proliferation of information-age technologies, however, these traditional military notions of environment and domain are changing. The information environment, for example, is defined as "[t]he aggregate of individuals, organizations, and systems that collect, process, disseminate, or act on information" (U.S. Joint Chiefs of Staff, 2010b), which are represented by the nodes across the traditional environments in Figure 1.4.

---

[4]   The Army is responsible for the land domain. The exception is the space domain, where the U.S. Air Force is the executive agent.

Figure 1.3
Traditional Military Domains

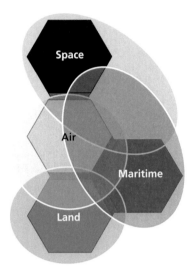

RAND *MG1113-1.3*

Figure 1.4
Information Environment

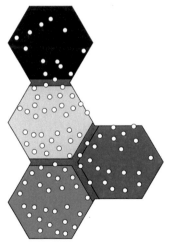

RAND *MG1113-1.4*

Cyberspace Defined

Cyberspace, itself, has become something of a portmanteau word—that is, it brings together two separate ideas into one cohesive concept.

Multiple interpretations are reflected in the many attempts to define cyberspace. Former Deputy Secretary of Defense Gordon England provided some guidance to DoD when he defined the term in a May 12, 2008, memorandum as follows:

> A global domain within the information environment consisting of the interdependent network of information technology infrastructures, including the Internet, telecommunications networks, computer systems, and embedded processors and controllers.

This definition is depicted in Figure 1.5, which includes the nodes and the connections (wired and wireless) between them that compose cyberspace.[5]

Figure 1.6 presents a less abstract picture of what constitutes cyberspace today, though we should point out that cyberspace is much more expansive than what is depicted in the figure.

According to Secretary England's definition, cyberspace has become another domain of warfare, but one that differs from the traditional domains in that it has both a physical and an informational component. The traditional forms of warfare—attack, defense, and exploitation—still occur, but information is the target, rather than people or materiel.

The quest for control of information has always been and always will be a part of conflict between warring parties (see Fitsanakis and Allen, 2009). Decades ago, the U.S. military's information infrastructure was contained primarily on paper and in human brains. Information technologies are clearly changing that. Orders of magnitude more information can be created, stored, sorted, and acted upon when that information is digitized. The real leap is occurring in terms of the abil-

---

[5]   Secretary England's definition does not include individuals and organizations, which are part of the information environment. Whether the omission is intentional or not, it makes the description of cyberspace more consistent with the description of the traditional physical domains.

**Figure 1.5
Notional Illustration of the
Interdependent IT Network**

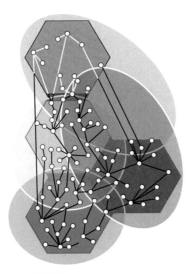

RAND *MG1113-1.5*

ity to disseminate information rapidly through large networks. The Internet is the obvious, if not the only, example. The emergence and growth of these networks constitute the birth and growth of cyberspace itself: a man-made, sprawling domain that continually changes and evolves over time and in space. Although its entirety is hard to visualize, it is an actual physical maneuver space where information can be attacked, defended, and exploited. As is the case with all communication infrastructure, adversaries will fight to control it.

A number of trends have accelerated the transformation of cyberspace into a new battlespace:

- the move toward digitized information (voice, video, and data)
- the miniaturization of computing and data-storage devices that carry digitized information, coupled with low costs, that has fostered an explosion of increasingly networked digital devices

Figure 1.6
Cyberspace Today

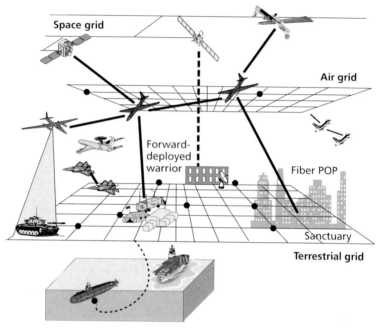

RAND *MG1113-1.6*

- the continued growth in wired and wireless networks and electronic systems, permitting access to systems that, until recently, may have been offline
- the combined decrease in cost, increase in speed, and standardization of interoperating electronic systems, which not only makes these systems more accessible to anyone, but also increases the potential for exploitation.

These and other trends enable any government or state to use technologies that were once available only to developed countries with large defense budgets, though it should be noted that these capabilities simultaneously increase the exposure of those countries.[6] Additionally,

---

[6]   Many relevant trends could be discussed in detail. For example, the increasingly popular approach to social networking enables individuals to remain online and active at all times.

individuals who were previously considered noncombatants can now join the battle and wage silent, electronic war. Finally, as information systems become more ubiquitous, the U.S. military's reliance on them increases apace. Today's modern economic, political, and military systems depend more than ever on information and instructions that are generated in cyberspace nodes and transmitted across a vast network. Such reliance makes cyberspace a natural arena for conflict.

Cyber-Electromagnetic/Cyber-Electronic Operations

Joint Publication (JP 1-02) defined the electromagnetic spectrum as the "range of frequencies of electromagnetic radiation from zero to infinity" (U.S. Joint Chiefs of Staff, 2010b). U.S. Army Training and Doctrine Command (TRADOC) Pamphlet 525-7-8 uses the phrase "cyber-electromagnetic contest" to highlight the overlap between cyberspace and the electromagnetic spectrum (TRADOC, 2010a). In this monograph, we refer to operations in cyberspace and the electromagnetic spectrum as *cyber-electromagnetic operations* (or *cyber-electronic operations*).

Purpose

This monograph argues the current doctrinal organization does not take into account trends in the cyber world and, further, that it imposes artificial boundaries that hamper the implementation of IO. Additionally, the personnel functional area that supports IO is not well understood and does not support effective IO implementation.

As a result, it argues that the Army's approach to IO needs to be redefined and reorganized and the personnel system tailored to support the new structure. A better organizing principle would be to separate psychological functions from technical ones. Such a division suggests that computer network operations (CNO) be merged with electronic

---

Developments such as cloud computing provide access to unanticipated amounts of computing power. More fundamental trends include the move from paper to digital media (for storing and documenting information), the use of digital communication for military command and control, and the reliance on software applications to control hardware devices and systems (i.e., digital control).

warfare (EW) and that personnel areas be established to support it. It also suggests that public affairs (PA) and military information support operations (formerly psychological operations, or PSYOP) have been artificially separated and need to be more explicitly integrated so that these efforts can be coordinated more effectively.

Approach

Our approach was to first clarify the taxonomy of IO by proposing definitions for key terms. Next, we identified the practical boundaries among types of IO tasks and, given those practical boundaries, assessed various organizational options. Finally, we reviewed how the Army personnel system supports the execution of IO; this monograph recommends some alternative personnel approaches. EW capabilities and tests were combined with electromagnetic spectrum operations (EMSO) capabilities and tasks to consider the overlap between the two areas.

How This Monograph Is Organized

The remainder of this monograph is organized as follows. Chapter Two provides background on two important concepts that are relevant to IO: the information environment and information warfare. In Chapter Three, we discuss the problem with the definition of IO in current doctrine. Chapter Four proposes changes to the definition. Chapter Five discusses EW, which is currently a functional area under IO. Chapter Six considers how PA, MISO, and strategic communication relate to IO. Chapter Seven examines how EW, cyber operations, and EMSO could be integrated. Chapter Eight proposes how PA and MISO could be better integrated. Chapter Nine presents our conclusions and recommendations.

The monograph includes six appendixes providing overviews of current doctrine, assessments of common tasks and areas of overlap, and a brief case study of IO organization in practice.

CHAPTER TWO

The Information Environment and Information Warfare

Before discussing IO in greater detail, it is first helpful to set the concept in a larger context. IO take place in the information environment and are a subset of information warfare (IW). This chapter further describes the information environment as introduced in Chapter One. It identifies its components and reviews past and proposed terms that are relevant to understanding how the information environment can and should be compartmentalized.

The Information Environment

The U.S. Department of Defense View of the Information Environment

As noted in Chapter One, the official DoD lexicon (JP 1-02) identifies *information environment* as the "aggregate of individuals, organizations, and systems that collect, process, disseminate, or act on information" (U.S. Joint Chiefs of Staff, 2010b). Joint doctrine (U.S. Joint Chiefs of Staff, 2007) provides further details on its make-up: "There is an electromagnetic spectrum portion of the information environment."[1] Thus, wired and wireless mediums fit in this landscape as well.

---

[1] In joint doctrine, *information environment* refers to the electromagnetic environment (EME). (See U.S. Joint Chiefs of Staff, 2007.)

Scholars, subject-matter experts, and DoD doctrine (U.S. Joint Chiefs of Staff, 2006b) describe three dimensions of the information environment: cognitive, informational, and physical dimensions. Wass De Czege (2008) describes these dimensions as mission tasks.

> It will be more important to pursue three ever-present, but practical, mission tasks. . . . One of these is to win the psychological contest with real and potential adversaries. Another is the need to keep the trust and confidence of home and allied publics while gaining the confidence and support of local publics. The third is winning the operational and strategic cognitive and technical "Info Age Applications" contest with real or potential adversaries.

A recent U.S. Army TRADOC capstone document (Pamphlet 525-3-0, 2009) divides the information environment into two areas:

> Because war remains fundamentally a contest of wills, prevailing in future armed conflict will require Army forces to exert a psychological and technical influence. Psychological influence efforts employ combinations of cooperative, persuasive, and coercive means to assist and support allies and partners, protect and reassure populations, and isolate and defeat enemies. Exerting technical influence entails protecting friendly information and communications and disrupting the enemy's ability to move and manipulate information.

Components of the Information Environment

As shown in Figure 2.1, the ways in which DoD describes cyberspace and social networks (e.g., Facebook) in JP 1-02 fit within broader construct of the the information environment (U.S. Joint Chiefs of Staff, 2010b). Although not shown, doctrinally defined areas, like EW, CNO, MISO (formerly PSYOP), and EMSO, are also included. These areas fit into the concepts of cyberspace, the information environment, or both.[2] With respect to IO, the presence of social networks in the

---

[2] According to Elder (2010), "EW, CNO, PSYOP, and EMSO are best defined as operations with effects in the information environment; the activities themselves are actually conducted in the cognitive (social), logical (cyber), or physical realms."

Figure 2.1
Cyberspace and Social Networks in the Information Environment

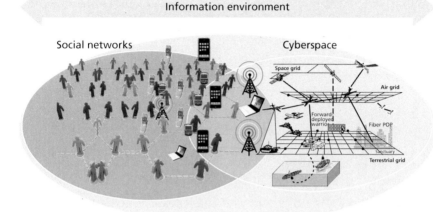

The information environment is increasingly digital and wireless.

RAND *MG1113-2.1*

information environment and the environment's overlap with cyber-space are important developments. As noted by LTG Michael Vane, "Army forces operate in and among human populations, facing hybrid threats that are innovative, networked, and technologically-savvy" (TRADOC, 2010a, p. i). Internet-assisted social networking is now a part of the operational environment, as events in Moldova[3] and Iran[4] (and even Pittsburgh)[5] have made clear. They are a growing venue for developing influence.

---

[3] Cell phones and text messaging are believed to have played a crucial role in fostering the so-called Orange Revolution in the Ukraine. Twitter is credited with making these protests widespread and successful (e.g., flash mobs). Ultimately, the protests forced a recount of the general election. See Morozov (2009), Goldstein (2007), and Stack (2009).

[4] During Iran's so-called Twitter revolution, it was reported that well-developed Twitter lists showed a constant stream of situational updates and links to photos and videos, all of which painted a portrait of the developing turmoil. According to news reports, when the Iranian regime started taking down these sources, the so-called e-dissidents shifted to email. (See "Iran's Twitter Revolution," 2009.)

[5] During a recent G20 meeting, protesters in Pittsburgh leveraged Twitter. For example, Elliot Madison, an activist in New York City, used Twitter to disseminate an order-to-

Information Warfare

A Definition

Information warfare is not currently defined in joint or Army doctrine, but it *is* a term found in past doctrine (mid-1990s; see AFDD 5, 1996, and CJCSI 3201.01, 1996), in the militaries of other countries, and in job descriptions in other U.S. services. For example, the Navy now has an information warfare officer position, which it advertises as involving "attacking, defending and exploiting networks to capitalize on vulnerabilities in the information environment" (U.S. Navy, undated). For our purposes, we define IW as follows:

> *Information warfare is conflict between two or more groups in the information environment.*[6]

Debate over the Term

Debate over the meaning and even significance of the term *information warfare* extends back at least 15 years. As noted by Buchan (1996):

> various organizations are defining it differently and emphasizing different aspects of the problem. Although groping for an acceptable definition appears to have absorbed an inordinate amount of the defense community's attention in recent months, ambiguities

---

disperse message from the Pittsburgh police during the protests. Reportedly, police raided Madison's hotel room, and, one week later, his home was raided by FBI agents. Police reports claim that Madison and a co-defendant used computers and a radio scanner to track police movements and then passed that information to protesters using cell phones and Twitter. Madison is reportedly being charged with hindering apprehension or prosecution, criminal use of a communication facility, and possession of instruments of crime (Democracy Now! 2009; Electronic Frontier Foundation, 2009; Goodman, 2009).

[6] See Buchan (1996). Dan Kuehl at the National Defense University defines IW as "Military offensive and defensive actions to control/exploit the environment" (various briefings); U.S. Joint Chiefs of Staff (1995) notes that "IW focuses on affecting an adversary's information environment while defending our own." CJCSI 3210.01 (1996) defined information warfare as follows: Actions taken to achieve information superiority by affecting adversary information, information-based processes, information systems, and computer based networks while defending one's own information, information-based processes, information systems and computer-based networks."

still remain. The basic point of contention seems to be the scope of information warfare: whether it is basically limited to:

[a] conducting or defending against electronic attacks on computers and related information systems or

[b] whether it also includes the whole spectrum of possibilities for using information effectively in warfare and denying enemies the same capability.

Functional Areas That Compose Information Warfare

Many areas are integral to IW, including EW, EMSO, CNO, and military deception (MILDEC), as well as MISO (formerly PSYOP).[7] Table 2.1 presents a more complete list. As we discuss later, operations security (OPSEC) can be made a part of mission command. Some MILDEC can be considered EW (e.g., the use of expendables and flares).[8] Today's Army organization and doctrine treat many of the functional areas as separately operated, compartmentalized capabilities. We argue, however, that the past decade has brought about a different landscape that renders many of these functional areas, and the aforementioned scopes, inseparable with regard to achieving desired, coordinated effects.[9]

Figure 2.2 shows the taxonomy of the information environment, IW, and IO. It is not intended to suggest that IO are the only types of IW in the information environment. As Table 2.1 indicates, network operations and EMSO, to name just two, also occur there. The purpose of the figure is to illustrate the relationship among the three concepts.

---

[7] DoD's 2010 decision to re-label PSYOP as MISO was a move to more accurately reflect the breadth of such operations and to distance the field from the implications that the name *PSYOP* had gained over the years.

[8] Aspects of EW include erroneous signals or information targeting machines, which may, in turn, deceive decisionmaking authorities and systems operators (e.g., smart jamming and expandable decoys are designed to create inaccurate operational pictures). Such effects depend on both the quality of the technology and the expertise of the operators.

[9] We recognize that other military functional areas have effects, by either "blowing things up" or generating signals or messages in some other way; actions speak louder than words.

Table 2.1
Doctrinally Defined Functional Areas

Functional Area	Army Field Manual	Selected Subareas, Divisions, and Activities
Electronic warfare (EW)	FM 3-36 (2009)	Electronic attack (EA), electronic protect (EP), electronic warfare support, spectrum management and control
Computer network operations (CNO)	FM 3-13 (2003)	Computer network attack (CNA), computer network exploit (CNE), computer network defense (CND)
Network operations	FM 6-02.71 (2009)	Information assurance (IA)
Electromagnetic spectrum operations (EMSO)	FM 6-02.71 (2010)	Spectrum management, frequency assignment
Information operations (IO)	FM 3-13 (2003)	EW, CNO, PSYOP, MISO, OPSEC
Signals intelligence (SIGINT)	FM 2-0 (2010), FM 34-1	Gathering intelligence by intercepting signals
Military information support operations (formerly PSYOP)	FM 3-05.30 (2005), FM 3-13	Influencing emotions, motives, objective reasoning, and behavior
Public affairs operations	FM 46-1 (1997)	A focus on U.S. forces, populations, coordinating with MISO but remaining separate
Knowledge management	FM 6-01.1	Creating, organizing, applying, and transferring knowledge

The Air Force still defines IW as consisting of three elements: influence operations, EW, and CNO.

The Terms *Cyber-Electronic* and *Cyber-Electromagnetic*

As much as possible, we use the terms *cyber-electronic* (or *cyber-electromagnetic*) and *cyber-electronic operations*, as opposed to simply *cyber* or *cyber operations*. As Elder (2010) notes, this is a more useful way to characterize the larger system of interrelated and connected technologies.

Figure 2.2
Information Operations and Information Warfare in the
Information Environment

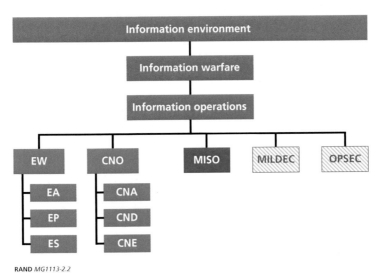

RAND *MG1113-2.2*

The Army's cyberspace concept capability plan (TRADOC, 2010a) discusses cyber-electromagnetic operations for the same reasons. Specifically, it refers to the "cyber-electromagnetic contest."

Wass de Czege (2008) argues for the use of the term *cyber-electronics* as well: "Cyber-electronics is a term I prefer over Cyberspace to cover the science that bounds and defines modern communications, including the Internet. Moreover, the character of modern operations is so shaped by these sciences, and the enabling capabilities that stem from them, that to not consider these a 'dimension' would be limiting." To an extent, the term *cyber-electronic,* or even *cyber-electromagnetic,* shows deference to an older definition of cyberspace (Elder, 2010). The older definition was included in the National Military Strategy for Cyberspace Operations, published in 2006 and recently declassified (U.S. Joint Chiefs of Staff, 2006b). That publication defines cyberspace as "a domain characterized by the use of electronics and electromagnetic spectrum to store, modify, and exchange data via networked systems and associated infrastructures."

CHAPTER THREE

The Problem with Information Operations

How Information Operations Are Defined

Current joint doctrine defines IO as

> the integrated employment, during military operations, of information-related capabilities in concert with other lines of operation to influence, disrupt, corrupt, or usurp the decision-making of adversaries and potential adversaries while protecting our own. (Gates, 2011)

How Information Operations Are Organized in the Army

Figure 3.1 is an expanded version of Figure 2.2 in Chapter Two. It depicts the doctrinal organization of IO in the Army today. As the figure shows, doctrinal IO emphasizes the integration of five "core capabilities" (though it also includes the integration of a number of "supporting" and "related" capabilities). The shading divides these five capability areas into two subcategories: those focused on content and those that are technologically enabled. MISO (formerly PSYOP), OPSEC, and MILDEC all focus on the content of a given message or action.[1] IO in these areas will succeed or fail depending on how well the content of the message melds with its purpose. For example,

---

[1] Certain functions of EW can be considered MILDEC. This includes the use of expendibles (e.g., flares) by vehicles (Hura, 2010). This should be (and likely is already) included in EW doctrine and/or corresponding tactics, techniques, and procedures.

Figure 3.1
Doctrinal Organization of IO

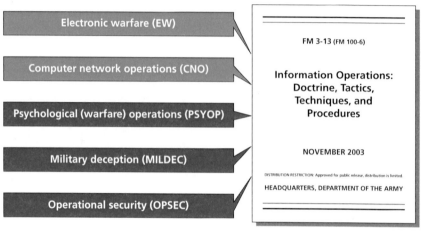

RAND *MG1113-3.1*

MISO efforts succeed when the message has been well tailored to the target audience, taking into account such issues as tribal associations, religious beliefs, cultural traits and history, and political context. EW and CNO, on the other hand, do not depend on the quality of the message. Indeed, there is no "message," per se, involved. These operations' success hinges on the technological quality of the equipment and the technical skill of the operators. For example, the success of CNA depends on how skillful the hackers are (e.g., whether they are good programmers, whether they understand the important computer languages) and the quality of the technology that supports them.

Another way of thinking about the divisions in the current IO organization is to consider them from the perspective of what they target. EW and CNO target machines. MISO, OPSEC, and MILDEC target people. Some MILDEC targets machines as part of EW (e.g., through the use of expendables or flares). Table 3.1 illustrates this division. The shaded cells indicate the five doctrinal IO areas.

Table 3.1
Targets for IO

Area	Primary Target
EW (ES, EP, EA)	Machines
MISO	People
PA	People
CNO (CNE, CNA, CND)	Machines
MILDEC	People
OPSEC	People and machines

Problems with the Current Definition

A great deal of confusion is associated with IO. This confusion has been well documented by subject-matter experts, practitioners, and academics alike. Consider the following article titles and quotes already published in reputable, relevant journals:

- *Why Warfighters Don't Understand Information Operations* (subtitle of an Army War College issue paper by Murphy, 2009).
- "There seems to be a lot of confusion in the Army as to the exact nature of information operations" (Rohm, 2008).
- "IO is a horrible term—it is at once everything and nothing. It can mean almost anything you want it to and is often used to mean very different things" (Beebe, 2009).
- "Army IO Is PSYOP: Influencing More with Less" (title of a *Military Review* article by Boyd, 2007).

Confusion stems from many sources: genuine ambiguity in the lexicon, both willful and unintentional misuse of the term, and both genuine misunderstanding and genuine disagreement about what IO is and what it ought to be.[2] Because there are genuine disputes regarding both the terminology and concepts of IO, resolution cannot be achieved with simple clarification. There are decisions to be made.

The current definition does little to clear up the confusion, due to its ambiguities, the fact that soldiers imagine it to (or want it to) mean something else, and the reality that IO *as actually practiced* deviates from that definition (Paul, 2008). A further discussion of the history and evolution of IO doctrine is presented in Appendix B.

Misunderstandings and Disagreements

According to Beebe (2009), who refers to the definition of IO that was in effect at the time of this research and prior to the January 2011 memorandum issued by Secretary of Defense Robert Gates,

> In the U.S., our IO doctrine is spelled out in JP 3-13: Information operations are "the integrated employment of electronic warfare (EW), computer network operations (CNO), psychological operations (PSYOP), military deception (MILDEC), and operations security (OPSEC), in concert with specified supporting and related capabilities, to influence, disrupt, corrupt or usurp adversarial human and automated decision making while protecting

---

[2]   As Professor Dennis Murphy of the U.S. Army War College notes in his 2009 issue paper (p. 1), "A review of current military and U.S. government information-related lexicon and definitions points out a very obvious flaw: this stuff is confusing . . . and in some cases, self-defeating." Similar concerns about the information lexicon have been raised by Paul (2008, 2009a, 2009b).

See Boyd (2007, p. 69) for a discussion of intentional misuse. See Rohm (2008) for discussion of both intentional and unintentional misuse. See Dominique (2010) and Greenmyer (2010) for discussions of unintentional misuse. The misuses most commonly cited are (1) a preference for referring to PSYOP as IO, because of either a misunderstanding or an intentional desire to avoid saying PSYOP, which has pejorative connotations, or (2) referring to IO as though it is a capability and produces products rather than an integrating function.

On the third point, Allen (2007, p. 17) notes that "a lot of the misunderstandings about IO occur simply because folks are used to one aspect of IO, especially if they came from a community that had its own cultural approach." Similarly, there are genuine disagreements about what IO should be and where it should go. Consider, for example, the 2009 draft revision of FM 3-13, *Information*, which was delayed and ultimately rejected due to disagreement over its content (Gould, 2009).

our own. Unfortunately, few people outside the IO community understand the definition—and even fewer (including many in the IO community) understand how these diverse disciplines are theoretically combined into an effective IO plan. At the service level, each branch understands the term differently—with more or less emphasis based on the individual service competency or viewpoint. For example, most soldiers think of IO as influence operations, and PSYOP is predominant. For the Air Force, it's mostly about CNO. . . . Primarily, when people use the term "information operations" to describe the adversaries' actions in the battlespace, they mean propaganda. They are talking about influence operations—not the integrated employment of EW, CNO, PSYOP, MILDEC and OPSEC.

Problems with the Current Organization

One problem with the current structure is that it creates a tendency to view IO through the lens of individual functional areas. Current doctrine emphasizes EW, CNO, MISO (formerly PSYOP), MILDEC, and OPSEC.[3] This emphasis inclines a reader toward conflating these various capabilities, and paying less attention to the possible integration of supporting or related capabilities or to value that might be added by capabilities outside of those listed.

A second, much more fundamental problem, as has been argued elsewhere, is that the current assemblage of core capabilities in IO (MISO, CNO, EW, MILDEC, OPSEC) conflates the "apples" of content and the "apple carts" of systems (Paul, 2008). Capabilities that generate content differ fundamentally from capabilities that affect systems. While there is an important interrelationship between systems and content (as the apples/apple-carts metaphor suggests, an empty apple cart is pretty useless, and it is hard to deliver a bunch of apples without something like a cart), they are of a wholly different character, follow different processes, and require different training and expertise.

---

[3]   CNO and cyber operations are synonymous terms and areas. In this monograph, CNO also refers to what many call cyber operations.

A third problem is that there is no MILDEC or OPSEC officer or noncommissioned officer specialty (it is a function), and deception is coordinated at many levels and across staff elements (i.e., all deception operations require broad input to be effective). This problem worsens at the lower echelons, where there are fewer and fewer dedicated IO soldiers and assets and less training, even though this is where much of IO is actually conducted (particularly in current operations).

A fourth problem is that IO planning, integration, and capability are different at different echelons. There are almost no integrators (if fire support officers are not included) at the battalion-and-lower echelons, even though these echelons will likely receive tactical MISO teams and civil affairs teams or other assets. All have to be coordinated as part of an integrated operations plan.

Table 2.1 in Chapter Two showed areas that are integral to information warfare. The Army's current organization and doctrine treat many of the functional areas as separately operated, compartmentalized capabilities. Whether such treatment was ever effective, it is not relevant today. We argue that the past decade has changed the landscape in such a way that renders many of these functional areas inseparable from one another with regard to achieving desired, coordinated effects. (See Appendix E for a discussion of the 1/25 Stryker Brigade Combat Team [SBCT], specifically.) Thus, any organizational scheme that does not recognize these interdependencies is flawed.

The Army has taken the first steps toward recognizing these interdependencies in its efforts to develop doctrine for cyber operations. This emerging doctrine acknowledges the convergence of mediums that is affecting the landscape. According to TRADOC Pamphlet 525-7-8 (2010a), the operational environment has been transformed by the "technologic convergence" of wired and wireless networks in general and computer and telecommunication networks in particular. TRADOC (2010a) asserts that this "technologic convergence of computer and telecommunication networks" is dramatically changing the operational environment and blurring the lines between the doctrinal areas of IO.

A direct implication of these convergence trends is that EA and CNA will soon operate on the same (merged) playing field, Similarly,

EP and CND will operate on the same playing field, as will ES and CNE. In short, EW (EA, EP, and ES) and CNO (CNA, CND, and CNE) are becoming increasingly comingled and could potentially share the same people, process, and technologies.[4] This convergence argues for new definitions, organizational boundaries, personnel specialties, and training.[5]

Lack of Common Vision for Information Operations

Several different visions of IO currently compete for acceptance, including some implied by the joint definition. Here, we characterize these visions and consider their merits. A future vision of IO will likely include elements from one or more of these views.

Vision 1: Base Case

Analyses of alternatives always consider the base case against which alternatives are to be compared. A caricature of the vision implied by the current joint definition is that IO is a poorly understood and vaguely bounded integrating function that coordinates disparate capabilities in pursuit of ambiguous objectives (see HQDA, 2003, and U.S. Joint Chiefs of Staff, 2007). It is flawed because it is an enumeration of a collection of capabilities.[6] It is not a compelling vision, and it makes clear the need for greater clarity.

---

[4] According to Elder (2010), "Although cyber and EW can achieve similar effects, the skills to conduct these activities are very different." Consolidation is always attractive because it suggests efficiency; however, improper consolidation can lead to loss of experience in critical capabilities (as occurred with EW in recent years).

[5] Emphasis must remain on coordinating with content, however, just as physical destruction should not be separated from its message. If a picture is worth a thousand words, then a JDAM is worth 10,000.

[6] The revised definition (Gates, 2011) mitigates this issue.

Vision 2: Information Operations as a Coordinating and Integrating Function

A better vision is one that makes the case for the capabilities that need to be integrated without resorting to a simple enumeration of key capabilities. A better vision projects[7] IO as a function to support the coordination and integration of various capabilities in pursuit of a range of information objectives (see HQDA, 2003, and U.S. Joint Chiefs of Staff, 2007). For this reason, some contemporary proponents want to do away with specific lists of capabilities (Murphy, 2009; Williams and Romanych, 2009) but retain a vision of IO as the practice of deconflicting and synergizing different capabilities to achieve effects that are greater than the sum of their parts. The revised joint definition (Gates, 2011) does exactly that.

Vision 3: Information Operations as Command-and-Control Warfare

IO grew out of command-and-control warfare (C2W) in the late 1990s, and some segment of the Army retains that initial vision. This was the result of a focus on commanders and their decision cycles. A 1997 version of FM 34-37, titled *Strategic, Departmental, and Operational IEW Operations*, defines IO as

> continuous military operations within the military information environment that enable, enhance, and protect the friendly force's ability to collect process and act on information to achieve and advantage across range full range of military operations. Information operations include interacting with the global information environment and exploiting or denying an adversary's information and decision capabilities. (HQDA, 1997a, Chapter 3)

IO as C2W is a narrow conception, focusing IO exclusively on affecting adversary information and information systems–based decisionmaking while protecting U.S. assets. IO seeks synergies between a tight set of capabilities for a focused set of objectives. This vision can be caricaturized as "whoever can keep their OODA [observe, orient, decide, act] loop spinning faster, wins" (Coran, 2004).

---

[7]   In current doctrine (FM 3-13 and JP 3-13.1).

Vision 4: Information Operations as Influence Operations

Under this vision, IO are military efforts to influence foreign populations or adversaries. This view seems to be the general consensus, anyway, given the way in which IO are discussed in Congress and by senior leaders.[8] Surely some of this view stems from the role such operations have played in Afghanistan and Iraq, focusing mostly on persuasion and influence through the MISO capability (only one of the traditional five pillars of doctrinal IO). Nonetheless, there is a growing body of thought that views influence as conceptually and operationally distinct from the more technical elements of traditional IO (notably EW and CNO) and asserts that the two should be divided.[9]

Vision 5: Information Operations as Advocacy

Because information and influence effects lie outside the bounds of traditional military training and thinking, one could argue that information needs an advocate or proponent as part of the commander's staff.[10] The information proponent would be available to predict the possible information effect of actions under consideration, remind the commander (and planners) to consider the cognitive implications of their planned actions, ensure awareness of various information-related capabilities and how they could contribute to operations, and serve as a source of advice regarding things informational.[11] Under this vision, that would be the role of IO.

---

[8] See Boyd (2007, p. 69) and Rohm (2008) for examples of senior leaders using IO to denote influence. See Ambinder (2010) and Gertz (2009) for examples of congressional presumption that IO is just influence.

[9] See, for example, the new "information tasks" in HQDA (2010b) or the apples/applecarts argument in Paul (2008).

[10] Most brigade staffs (but not battalion or lower) have an IO officer. Each senior staff officer is a principal adviser to the brigade commander. Whether the brigade commander decides to utilize IO and the advice of this officer is another matter. For example, the 3/2 SBCT downplayed IO, while the 1/25, discussed in Appendix E, elevated IO.

[11] Elder (2010) states, "Recognizing that adversaries and allies alike conduct their decision calculus based on their info and perceptions, as a minimum, it is important to ensure the information used in the calculus is accurate, and not biased against the US and our allies."

Vision 6: Information Operations as Everything

Another vision holds that, since everything generates information or affects the information environment in some way, everything is IO. We mention this position as a cautionary inclusion, or a risk: If IO is everything, then in practice, IO becomes nothing (Paschall, 2005; Armistead, 2004). If the concept that "IO is C2W" is derided as too narrow a vision for IO, then this is the polar opposite.

The visions described here cover a wide spectrum, from an integrating capability to a form of C2W to an area of emphasis for an advocate. This is not to say that any of the visions is necessarily wrong. Indeed, elements of several correctly characterize some aspect of IO. The problem is that none captures all the essential elements of IO. Furthermore, the fact that these visions differ so much illustrates that IO means different things to different people. Thus, it is unsurprising that implementation is uneven or contradictory at the operational level. The competing visions are also an argument for defining IO more clearly and developing a more rational organizational approach.

Visions 3 and 6 are shared today in many quarters and represent problems that exist with IO today. However, we believe that visions 4 and 5 are better and recommend moving toward a shared vision that incorporates aspects of both 4 and 5.

Information Operations as a Moving Target

Further complicating the situation is that the need for change in IO is recognized, and progress is under way as of this writing in both the joint community and the Army toward improving definitions, revising doctrine, clarifying concepts, and adjusting organizations. The 2011 memo "Strategic Communication and Information Operations in the DoD" by Secretary Gates, which formalized a new joint definition for information operations, is an example of progress in this area. We have endeavored to stay abreast of such movement, but other changes and advances have taken or are taking place at the time of publication. Undoubtedly, some important decisions will have been made, other important progress will have been occurred, and some of the recommendations presented here will have been overtaken by events.

However, challenges will remain. Debate within and surrounding the IO community runs hot and fierce, misunderstandings persist, and progress is often delayed by disagreements (such as the 2009 attempt to revise to FM 3-13, described in Appendix B). The information environment continues to evolve, adding new challenges.

CHAPTER FOUR

Redefining and Reorganizing Information Operations

This chapter attempts to construct a more coherent vision of IO that captures its essential functions. The chapter is organized as follows. First, we pose three key questions that need to be addressed to better define IO. Lengthy discussions follow that provide a range of possible answers for each question considered. At the end of the chapter, we propose a new definition of IO.

Key Questions and Answers to Guide a Redefinition of Information Operations

Our approach to defining IO was to pose three questions, the answers to which would help develop a coherent vision for IO:

1. Is the role of IO integration, advocacy, or a capability in its own right?
2. What is being integrated, advocated, or executed?
3. To what end?[1]

In the following sections, we offer possible answers, review various positions and views supporting some of those answers, and discuss the implications of the different answers. Table 4.1 summarizes the questions, options, our answers, and reasons.

---

[1] As noted by Elder (2010), "If the third question is asked first, its answer will inform the other two."

Table 4.1
Three Questions and Answers Toward a Vision of IO

Question	1. What Is the Role of IO?	2. What Is Being Integrated, Advocated, or Executed?	3. To What End?
Options	Integration, advisory, or a capability in its own right?	Five core and host related and supporting capabilities Unspecified broad activities Everything that is said and done Just information content Just information systems (information technology and electronics)	"Information superiority" Information engagement C2W Information protection OPSEC MILDEC Inform and influence activities Cyber-electronic activities
Answer	The role is mostly about integrating certain capabilities and advising the commander about them. An IO officer who has been previously trained and is proficient in one of the areas (e.g., MISO) could indeed provide that capability, but that capability has its own name and doctrine.	There is no right or wrong answer as long as the scope is manageable. We choose to focus in information content (messages).	There is no right or wrong answer. Historically, IO has origins in all of these options. Scope must be limited, however. We choose inform and influence activities.
Rationale	Historically, IO officers have not had the training and subsequent proficiency to carry out the individual capabilities.	In OIF and OEF, commanders seek IO to inform and influence audiences. There are other functions that can cover the other areas like C2W.	In OIF and OEF, commanders seek IO to inform and influence audiences. There are other functions that can cover the other areas like C2W.

NOTE: OEF = Operation Enduring Freedom. OIF = Operation Iraqi Freedom.

Discussion on the Role of IO

The first question is whether the role of IO is intended to be integration, advocacy, and/or a capability in its own right. The phrase "integrated employment of" at the beginning of the joint definition that was in place at the time of this research and prior to the 2011 Gates memo, the discussion of organizing an IO cell in the doctrine, and the training received by professional IO officers (Functional Area 30) have all led to IO being seen and practiced as an integrating function. As one IO officer put it, "We're like BASF. We don't make IO; we make IO better."[2] As another IO officer reminds us,

> The IO product is not a handbill or a news release but rather a synchronization matrix or a tool that ensures that the information capabilities or various elements related to IO are synchronized to achieve the commander's desired effects. (Dominique, 2010)

Confusion over the Terms *Operation* and *Integration*

While IO as defined in doctrine today is about integration, this point is often misunderstood. This is, in part, because IO is a misnomer and as a result, commanders expect that their IO officer is going to do something operational. But, they are not trained to do so.

An Analogy. A simple analogy illustrates the nature of the problem. Consider the fire support cell. Commanders know that they can take their intent or desired effects to the fire support cell and the fire support coordinator will identify the best assets to direct the needed fires against the appropriate targets. No commander imagines that the fire support coordinator is going to get in a Humvee, ride out to an artillery formation, and personally lay a gun. It is understood that the fire support coordinator is a coordinator and integrator, not a trigger-puller or operator. Why then, would anyone expect the officer in charge of *integrating* IO capabilities to produce leaflets (a MISO function) or generate a press release (a PA function)? But that is exactly what happens in the field today.

---

[2] Anonymous interview with the authors.

The Need for Advocacy

Another possible answer (or part of an answer) is for IO to advocate for information and cognitive effects. Currently, FA30s are trained integrators; they are not required to have a particular level of mastery in any of the five core IO capabilities. Even if they have a background in one of those capabilities, that is usually all: *one* of those capabilities. While, currently, IO officers cannot execute any of the capabilities they integrate, they do know that all of the capabilities can be important, they have some knowledge about each, and they know how to look for synergies among those capabilities and with the broader capabilities of the joint force.

On Generating Effects. Because (as discussed in greater detail later) Army officers are trained in combined arms operations but are not currently well versed in the nuances of generating effects in the information, cognitive, or cyber domains, there is a gap in thinking about these effects.[3] Until information effects are inculcated in the culture and thinking of the officer corps, the argument goes, the Army needs specialist information advocates who are trained to think about these effects and whose sole job is to ensure that other elements of the staff have both a reminder to consider these effects and a conduit for access to needed expertise.

IO as a Distinct Capability. As a final consideration to this first question, given the general tendency to want to operationalize IO, why not just do so? Cut out the middleman or specialist integrator, combine the personnel who comprise or generate the capabilities needed in an IO cell; have the cell chief report to the G-3, the chief of staff, or directly to the commander; and then have that cell chief go back and generate information effects.[4] Many commanders may not feel that they want or

---

[3]   According to Elder (2010), "Army officers are trained to understand 'DIME on PMESII' but might benefit from opportunities to apply capabilities to influence the social and informational aspects of PMESII." DIME refers to diplomatic, informational, military, and economic. PMESII refers to political, military, economic, social, information, and infrastructure.

[4]   As Elder (2010) notes, "There is a great need to improve integration of IO capabilities at the operational level; however, at the tactical (or functional level), (to use an analogy) it will be better to let the carpenters build the house and cabinet maker build cabinets, even though they use similar tools." Physical IO assets are assigned to the functional level (e.g.,

need an integrator for their information effects but that they do want or need more capability to generate information products.

What Is Being Integrated, Advocated, or Executed?

A second key question: What should be under the IO umbrella and the relationship implied between IO and those capabilities?

Key Capabilities Need to Be Included

The authors of this monograph have various backgrounds. All consider themselves well steeped in IO. Some are engineers and have been working on cyber operations, EW, EMSO, and other spectrum-related topics for years (the "technical guys"). Others include sociologists who have spent years working on topics related to influence, shaping, MISO, and strategic communication (the "content guys"). While both groups have considerable expertise in their areas with regard to IO, they regularly learn things from each other. What this means is that the challenge facing an individual who is supposed to master both content and systems, whether as integrator, advocate, or executor, is formidable.

The division between the message (content) and the medium (information technology and electronics) is well captured in the two information tasks specified in the 2010 draft rewrite of FM 3-0: inform and influence activities and cyber-electromagnetic activities.[5] We address these tasks in greater detail later.

Note, however, that while the information tasks establish a clear division between influence and cyber-electromagnetic activities, both

---

tactical MISO, civil affairs teams, combat camera, mobile public affairs detachments). Most importantly, the average soldier, in contact with the population or the adversary, is a very important element of the IO campaign and must understand the information elements of his or her actions—not just themes and messages, although those are important as well). A significant amount of influence is carried out at the tactical level.

[5]   Elder (2010) notes, "Cyber-electromagnetic activities as a descriptor is a good way to combine logical and physical aspects of IO and differentiate technical from cognitive (inform/influence); however, this is not one technical skill set: engineering (hardware) competencies are very different from computer science competencies."

are still information tasks. Should IO include just one, or both? If just one, which? And what happens to the other?

Avoid Long Lists of Capabilities to Define IO

Since the 2003 IO Roadmap, IO has doctrinally had five core capabilities (PSYOP, EW, CNO, MILDEC, and OPSEC) and a varying cast of supporting and related capabilities, including, significantly, civil-military operations, PA, military support to public diplomacy, combat camera, and physical destruction. Current discussions of this subject avoid listing specific capabilities (e.g., Murphy, 2009; Williams and Romanych, 2009; Kuehl, 2009; Emery, 2008),), as does the new joint definition for IO, presented at the beginning of Chapter Three (Gates, 2011) and the latest proposed revisions of FM 3-13 seen by the authors.

Advantage and Disadvantages of Listing Capabilities. Listing specific capabilities has benefits and drawbacks. The benefit of listing specific capabilities is that it makes clear who is in the tent, who is out, and exactly which capabilities must be integrated, advocated, or executed as part of IO—at least in principle. In current practice, the integrative relationships and lines of authority to the "related" and "supporting" capabilities have never been completely clear.

The downside to listing specific capabilities is it excuses all other functions and capabilities from being concerned with IO and provides a rationale to ignore an insistent IO officer: "I'm not one of *your* capabilities, so leave me alone!" Capabilities not specified (or not specified as core) may in fact be very important to particular information operations. In the inform-influence-persuade mission set, for example, one of the most important lessons learned in recent operations is the communicative value of *actions*, including both maneuvers and fires (Helmus, Paul, and Glenn, 2007). If IO is to include the inform-influence-persuade mission (open for discussion in the next section), then it should not be limited to just PA, MISO, and CMO; it needs to include at least an integrative relationship with all capabilities that communicate or influence. Trying to list all of those capabilities explicitly might well be a fool's errand. Indeed, the revised joint definition of IO (Gates, 2011) refers simply to "other lines of operation."

The Army War College's Professor Dennis Murphy argues that the current definition of IO can be much improved and focused by removing the reference to specific capabilities, leaving it open to any capabilities:

> By explicitly excluding a laundry list of capabilities, the definition is no longer self-limiting since the tools available are now constrained only by the imagination of the commander and his staff. While it may not be about everything you do, it certainly can be about anything you can do to achieve the desired information effects in support of the military operation, to include physical attack, i.e. actions. (Murphy, 2009)

Moving in the same direction as those who argue against specifying capabilities but taking it even further are those who assert that everything is an information operation.

Arguments Can Be Made That IO Is Everything

> To an extent, every operation is an Information Operation. Every patrol, every battle, every discussion is a chance to persuade the population to support the government. (Cummings and Cummings, 2009)

While this view risks taking things too far (remember, if IO is everything, IO is nothing), it contains an important truism: All actions and utterances of the joint force can send messages or signal allies, adversaries, or populations. While it may not be meaningful to call everything the joint force does an *information operation*, all of these capabilities can be integrated for both traditional combined arms effects and information/cognitive effects, and an information advocate could help relevant staffs remember that all activities have information consequences.

Discussion of Possible Ends to Be Sought

Now we turn to the third question: To what end? We are not the first to address this question. While the very first version of JP 3-13, *Information Operations,* listed five IO effects (destroy, degrade, deny, disrupt, and delay), between its publication in 1998 and 2004, joint and service doctrine and IO concept papers listed a total of 44 effects that IO should produce (Allen, 2007). These effects are listed in alphabetical order in Table 4.2. Note that most of these so-called effects are actually tasks (for instance, destroy is a tactical task to physically render an enemy force ineffective, but destroy would be an effect). Most of these terms can also be used to refer to influence.

Table 4.2
Sample IO Desired Effects Compiled from Various Sources

Access	Diminish	Mislead
Cascading network failure	Dislocate	Negate
Control	Disrupt	Neutralize
Coordination failure	Distract	Operational failure
Create information vacuum	Divert	Paralysis
Decapitate	Exploit	Penetrate
Deceive	Expose	Prevent
Decision paralysis	Halt	Protect
Defeat	Harass	Read
Degrade	Influence	Safeguard
Delay	Inform	Shape
Deny	Interrupt	Shock
Destroy	Lose confidence in information	Stimulate
Desynchronize	Lose confidence in network	Stop
Deter	Manipulate	

SOURCE: Allen, 2007, p. 38.

The List of Relevant Capabilities Remains Too Long to Be Useful

The long list of desired effects in Table 4.2 is a tall order for IO. We do not propose that this collection be used as a shopping list. It illustrates the wide range of effects that could be desired of IO.

The set of tasks out of which IO grew (the aforementioned five Ds, destroy, degrade, deny, disrupt, and delay) came directly from the concept's genesis in C2W. This is a narrow conception of IO, focused on affecting adversary decisionmaking and protecting friendly decision-making. Should IO *only* be about C2W? Should the concept emphasize some other narrowly defined set of tasks to the exclusion of a broader set of goals?

Pros and Cons of a Narrow Set of Ends

There are advantages to a narrow set of ends. It makes it easier to "stay in your lane" and makes it more straightforward to leave capabilities in their traditional functional organizations and staffing relationships, calling on them only when needed for the narrow set of objectives.

There are potential problems with a narrow set of objectives, too. With a broad name like "information operations," there will be an expectation that a wide range of ends can be pursued with those operations. Further, if IO have only a narrow set of ends, when the joint force needs to pursue other information-related ends to achieve operational or strategic objectives, IO may not be prepared to integrate, advocate, or execute as required.

Although Table 4.2 lists 44 alternatives for desired IO effects, contemporary discourse on the subject aggregates these effects into a smaller number of sets of possible ends.

Information Tasks as Potential Ends

FM 3-0 contains "information tasks." The 2008 revision contained five information tasks: information engagement, C2W, information protection, operations security, and MILDEC (HQDA, 2008a).[6] The 2010 draft revision of FM 3-0 contains only two information tasks: conduct inform and influence activities, and conduct cyber/EM activi-

---

[6] Tasks are conducted to achieve ends.

ties. These two sets of information tasks imply very different sets of ends and very different ways of talking about them. Although there are five information tasks in the 2008 revision, they collectively imply a much narrower set of objectives.

Neither set of information tasks is particularly specific about the ends that those tasks can be used to pursue. The canceled draft of FM 3-13 is much clearer about ends, specifying three challenges that commanders will face in full-spectrum operations:

> one, to maintain the trust and confidence of home and allied publics while gaining the confidence and support of local publics and actors; two, to win the psychological contest of wills with adversaries or potential adversaries; and three, to win the contest for use of information technology and the electromagnetic spectrum. (Whisenhunt, 2009, p. 29)

Influence as an End

In contemporary operations, as one Army officer put it, "IO is PSYOP" (Boyd, 2007). For many, IO is (and should be) fundamentally about influence, at the expense of other possible ends. Recent thinking on IO in the Australian army "views retention of the term 'information operations' as non-critical; however, retention of the underlying concept of 'influence' is seen as paramount" (Nicholas, 2008, p. 39). A 2009 *Military Review* article rejects concerns that IO should support technical capabilities, suggesting that experts in the Army's Network and Space Operations and Forces Development Signal Corps can provide far better support and that removing technical concerns would "better allow IO to concentrate on influence operations" (Richter, 2009, p. 110).

Command-and-Control Warfare

One possible end is C2W. C2W is exclusively focused on adversaries and on U.S. forces. Because of its roots in C2W, IO doctrine retains significant residual focus on adversaries to the exclusion of other possible targets.

A Proposed Definition of Information Operations

Table 4.1 summarizes our answers to the questions motivating this chapter. In this section, we propose a definition of IO based on the answers to the questions enumerated in Table 4.1.

Proposed Definition

Our proposed definition is predicated on the acceptance of many of the arguments and recommendations advanced in this monograph. Should one reach different conclusions or prefer different visions or courses of action, this definition would not suit. Specifically, this definition embraces the separation of the technical from the psychological and proposes to apply the term *information operations* solely to the latter. Our suggested definition is as follows:

> *Information operations are efforts to inform, influence, or persuade selected audiences through actions, utterances, signals, and messages.*

This definition and the associated vision have several notable characteristics. First, this definition separates the "apples" of information content from the "apple carts" of information systems and retains the term *information operations* to refer to the former, exclusively. Under this definition, IO includes only efforts to inform, influence, or persuade. "Technical" information capabilities and objectives (such as EW and CNO) are divorced from IO (and are discussed later in this monograph).

Second, this definition implies that IO is an integrating function, but one that integrates not just across the capabilities that generate messages and images (MISO, PA, combat camera), and not just across the doers of good deeds (civil-military operations), but across *all* actions, utterances, signals, and messages. This is explicit recognition that all actions communicate and that the possible signals generated by friendly force actions (including maneuver and fires) must be considered, planned for, and coordinated to realize maximum effectiveness in influence.

Third, this definition includes advocacy. It is not enough to define IO in this way, charge an IO officer with integrating all these capabilities, and let things run their course. IO becomes the responsibility of the commander, who must specify desired cognitive end states. The IO officer must have a position on the staff that allows direct access to the commander and the opportunity to advocate for the consideration of influence effects in operations and relevant staff sections.

IIO and ITO: Compartmentalizing the Definition

It is not necessary to redefine IO per se. The definition can be broken up into two groupings of functional areas. The first grouping is for those functional areas that have to do with informing, influencing, and persuading target audiences. These include the content-focused functional areas shown in Figure 3.1. The second grouping contains the rest (e.g., the technology-focused areas in Figure 3.1).

The first group is named inform and influence operations (IIO), and we can simply define IIO as we did earlier: *efforts to inform, influence, or persuade selected audiences through actions, utterances, signals, and messages.* Alternately, we refer to the functional areas in this grouping as areas contributing to effects within the psychological realm.

The second group is named information technical operations (ITO) and includes the technically focused functional areas labeled as such in Figure 3.1 in Chapter Three. Alternately, we refer to the functional areas in this grouping as areas within the technical realm. This is shown more clearly in Table 4.3.

Table 4.3
Compartmentalizing the Definition into Two Groupings or Realms

Category	Psychological Realm	Technical Realm
Functional areas, subareas, defined in existing doctrine	MISO, public affairs (PA), aspects of MILDEC	Electronic attack (EA), electronic protect (EP), electronic support (ES), computer network attack (CNA), computer network exploit (CNE), signals intelligence, electromagnetic spectrum operations (EMSO), information assurance, operating and maintaining networks (network operations), aspects of MILDEC
Target	People	Machines
Alternate name	Inform and influence operations (IIO)	Information technical operations (ITO)

CHAPTER FIVE

How Electronic Warfare Overlaps with Other Areas

Our proposed definition concerns one of the problems identified in how IO is characterized today. The next issue is how IO should be organized. This chapter takes up the issue of EW and how it overlaps with other key areas, including EMSO, SIGINT, and cyber operations.

Analysis of Electronic Warfare and Electromagnetic Spectrum Operations

We analyzed a number of EW and EMSO tasks and assessed their commonalities and overlap. Specifically, we assessed the degree of overlap in these tasks by comparing a list of EMSO tasks to a list of EW tasks. The *Electronic Warfare Capability-Based Assessment* (EW CBA), issued by the TRADOC Analysis Center, provided a list of EW tasks (TRAC, 2009). The Signal Center at Fort Gordon is currently conducting a CBA of EMSO. From this effort, we obtained a list of EMSO tasks (U.S. Army Signal Center, 2010).

The EW CBA characterized the basic EW capabilities, tasks, conditions, and standards necessary to support Army warfighting functions in the 2015–2024 time frame. The CBA produced 29 tasks: seven related to EA, six to EP, three to ES, and 13 to EW integration (EWI). A complete list appears in Appendix D.

The ongoing EMSO CBA produced 15 EMSO tasks. The complete list of the EMSO tasks also appears in Appendix D. We binned each of the 15 EMSO tasks into one of the four components of EMSO: spectrum management, frequency assignment, host-nation coordination, and policy implementation. Eleven of the 15 EMSO tasks were related to spectrum management. Three tasks were related to frequency assignment, and one to host-nation coordination. Of the EMSO tasks produced by the EMSO CBA, no EMSO task related to policy implementation.

We compared these two lists, asking the questions, "Is this EMSO task involved in accomplishing this EW task?" and "Is this EW task involved in accomplishing this EMSO task?" An affirmative answer to either question indicated that the two tasks overlapped.

Overall, we identified a total of 435 possible overlaps between 15 EMSO tasks and 29 EW tasks. Further analysis revealed 106 overlaps (24 percent). A more focused analysis between each EW component and EMSO task revealed that some EW tasks shared more overlap with EMSO tasks than others did (see Table 5.1).

The top three EMSO tasks that overlapped with the greatest number of EW tasks were as follows:

- Monitor and use spectrum common operating picture (COP) information in support of full-spectrum operations.
- Utilize host-nation comments in the spectrum nomination and assignment process.
- Provide EME information in either a networked or stand-alone mode (build EME COP).

The top three EW tasks that overlapped with the greatest number of EMSO tasks were as follows:

- Protect friendly personnel, equipment, systems, information and facilities from adverse EW effects.
- Protect the use of the electromagnetic spectrum (EMS), including spectrum management and radio frequency (RF) deconfliction.
- Coordinate and modify emission-control measures.

As the list of EW tasks shows, EP, electronic support (ES), and EWI are important EW tasks that support and enable EA tasks. Similarly, EMSO also supports and enables EA tasks. Furthermore,

Table 5.1
Overlapping EW and EMSO Tasks

Task Overlap	Number	Percentage
EP/EMSO	50/90	56
ES/EMSO	5/45	11
EW/EMSO	51/159	26

FM 3-36, titled *Electronic Warfare in Operations* (HQDA, 2009), asserts that spectrum management is a part of EW operations. It lists spectrum management as one of the activities related to EP and describes it as follows: "electromagnetic spectrum management is planning, coordinating, and managing joint use of the electromagnetic spectrum through operational, engineering, and administrative procedures." Consistent with the aforementioned doctrines, our results indicate that many of the EP, ES, and EWI tasks do overlap with EMSO tasks, suggesting that a reallocation of resources and efforts to equip, train, and execute these overlapping tasks may improve operational efficiency.

Relationship of Intelligence Capabilities to Electronic Warfare and Cyberspace Operations

Signals Intelligence, Electronic Warfare, and Electromagnetic Spectrum Operations

SIGINT includes significant structure (in both DoD and supporting civilian agencies) and is squarely aligned under intelligence. SIGINT has some commonality and overlap with cyber operations, EW, and EMSO.

Electronic Warfare and Signals Intelligence

It is noteworthy that the distinction between aspects of EW and intelligence can be subtle. According to JP 3-13.1,

The distinction between intelligence and ES is determined by who tasks or controls the collection assets, what they are tasked to provide, and for what purpose they are tasked. ES is achieved by assets tasked or controlled by an operational commander. The purpose of ES tasking is immediate threat recognition, targeting, planning and conduct of future operations, and other tactical actions such as threat avoidance and homing. However, the same assets and resources that are tasked with ES can simultaneously collect intelligence that meets other collection requirements. (U.S. Joint Chiefs of Staff, 2007)

Relationship Between Electronic Warfare and Cyber Operations

In this section, we consider a three-interconnected-dimensional framework proposed by TRADOC as a way forward for cyber operations, EW, and IO. We consider it as a system and, applying the reasoning of system architecture, consider the advantages of consolidating some elements and disaggregating others to improve performance. Using data from the TRADOC Combined Arms Center and cyber operations/EM contest CBA sources, as well as interviews with personnel from U.S. Army Communications-Electronics Research, Development and Engineering Center, Intelligence and Information Warfare Directorate (CERDEC I2WD), we conclude that this will come about if the cyberwarfare and EW elements are consolidated, because their technologies, infrastructure, operational use, and personnel training are merging and they will benefit from common doctrines. This conclusion is supported by the activities and plans of the Army in Afghanistan, the Navy, and the Air Force.

Approach: A System Consideration of Cyber Operations

Figure 5.1 can be considered a system in that it is a collection of elements that function together to provide a function—in this case, cyber operations—that is not provided by the constituent elements themselves (Rechtin, 1991).

Figure 5.1
Cyber Operations and Enablers

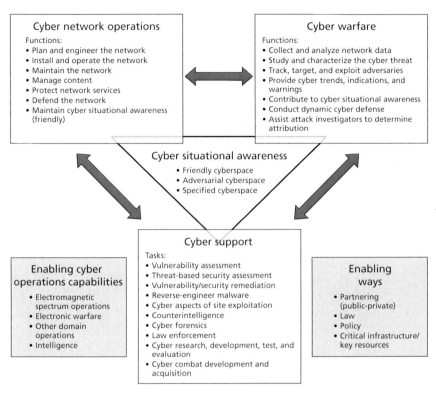

SOURCE: TRADOC, 2010a, p. 19.

RAND *MG1113-5.1*

These elements instruct on the importance of identifying system boundaries that separate what is inside from what is outside the system, allowing us to partition (i.e., aggregate or consolidate and disaggregate) the system in ways that facilitate analysis. The adversaries' cyber operations and (constituent elements thereof) would be outside this model, as would other influences, such as weather, which can affect EMSO. Partitioning means aggregating the interior elements into subsets for a useful purpose, and in one sense, we have done that by creating the box and triangle structure to facilitate consideration of the major subsystems that they define. The partitioned subsets constitute subsystems that interact with one another to varying degrees. At one extreme, there is no interaction between any of the partitioned subsystems; they are self-contained stovepipes that serve the purpose of the system in their own unique ways. However, if they did interact, the system serves no purpose and should be replaced by its noninteracting subsets. At the other extreme, there are many subsystems—in the most extreme case, each contains only one element—and each subsystem interacts with every other subsystem, creating a somewhat chaotic dynamic. Paraphrasing Rechtin (1991), relationships among the elements of a system are what gives a system its greatest value; choosing the appropriate partitioning of a system into subsystems is critical to the design of the system, and in partitioning a system, one must choose the subsystems whose interior elements are strongly dependent on one another and largely—but not completely—independent of elements exterior to them that constitute the interior elements of the other subsystems.

This reasoning allows us to imagine partitioning cyber and EW differently, to consider which elements—which in this case represent functions, organizations, and technologies—should be in the same subsystem because they are strongly interdependent,[1] or at least related.

Cyberwar and EW are both means of using EMS technology to attack and defend against the enemy. Increasingly their means are digitally enabled. These similarities make them candidates for consolidation as explored further here.

The current Army cyber/electromagnetic contest CBA (U.S. Army, 2010) has issued definitions that incorporate EW as an element within cyber attack, specifically.

---

[1] Interdependence could be measured using doctrine, organization, training, materiel, leadership and education, personnel, and facilities (DOTMLPF) as a metric. If two elements employ similar technology, they are M-interdependent, in that both are dependent on the future success of that technology. If two elements use similar doctrine, they are D-interdependent, in that both are likely to be used if that type of doctrine is chosen. If two elements are organizationally compatible and interactive, or posses this potential, they are O-interdependent. The stronger the interdependencies across DOTMLPF, the better the reason to minimize or eliminate any boundaries that might otherwise be constructed between them. It is the placement of boundaries (consolidation) and the elimination of boundaries (disaggregation) as they concern the elements of cyber operations that is being explored in this section.

In the doctrine, a cyber attack is defined as "actions that combine computer network attack (CNA, an element of CyberWar) with other enabling capabilities (e.g., electronic attack—EA, an element of EW, physical attack; etc.) to deny or manipulate information and/or infrastructure in cyberspace." A contemporaneous white paper that addresses the CBA further defines EW as "refer[ing] to any military action involving the use of EM energy to control the EMS or to attack the adversary" (TRADOC Combined Arms Center, 2010). Generically, this implies that cyberwarfare is included in EW, as cyberwarfare would satisfy this definition, too. The white paper provides additional commentary that reinforces the notion of the convergence of cyberwarfare and EW by highlighting the

> widespread technological convergence between computers, communications, electronic devices and sensors—convergence at both the device level and the supporting infrastructure level—the iPhone has all of these functions. Over time, the infrastructures used for cyber, EW and EMSO will be come indistinguishable—technological convergence is enabling our network assets to become our EW assets, and vice versa. (TRADOC Combined Arms Center, 2010)

It adds,

> These elements result in operational convergence. CyberOps and EMS ops are increasingly drawing on the same capabilities, e.g. both rely on signals intelligence assets and spectrum managers. Inter-dependence is such that CyberOps is essential for integrated EW and EMSO. EW capabilities have utility for attack and defense of platforms, systems and networks. These capabilities are being employed in concert (in very sophisticated combinations) to attain necessary objectives. (TRADOC Combined Arms Center, 2010)[2]

---

[2] According to Elder (2010), "The convergence in operational terms is real; however, at the functional level, very different education and experience are associated with hardware systems (engineering), software (computer science), RF, optical, and other electromagnetic engineering specialties, and systems engineering (typically involved in the integration). The

To explore this point further, RAND visited CERDEC I2WD, a provider of EW, cyber, and SIGINT materiel (Axelband, 2010). CERDEC provided its view of the boundaries between IO, cyber operations, EW, and SIGINT in the context of likely responses to CBAs. The output of a CBA is a statement of functional capability gaps and materiel and nonmateriel means of closing them. The resulting documents, if a materiel development path is chosen, lead to an initial capabilities description, a materiel development decision, and an analysis of alternatives. These form the basis for a technology development program. During the technical development phase, a capabilities description document will be issued and, with the successful conclusion of the technology development program, will lead to an engineering and manufacturing development program of record. A key question was whether, given materiel solution requirements from a CBA or any of its following documents, the equipment developed by I2WD would be necessarily restricted to use only for cyber operations, IO, EW, or SIGINT in its functional approach. Recall that functional and top-level system requirements are often generic in that they do not dictate the particulars of a materiel development. For example, such a generic requirement might involve detection, identification, tracking, and killing a class of targets by nonkinetic means under certain conditions.

The answer was no. CERDEC would be limited only by technology (cost, schedule, risk, utility, and so on, were, of course, important considerations), and this would lead to the use of technology that provided the basis for a system that could best satisfy closing the functional capability gaps. It could envision that there might be circumstances in which satisfying the functional requirements of an EW CBA, a SIGINT CBA, or a cyber operations CBA could each lead to the same or similar technologies and techniques. For example, the functional requirements of an EW CBA could be met by a combination of what traditionally has been called EW, cyber operations, or SIGINT technologies and techniques. This would not happen if such a combination did not provide the best system solution or if, in addition to the functional require-

---

danger is that resource managers might attempt to extend the useful operational-level convergence to the functional level (as was done with IO before)."

ments, there were restrictions stipulating the use of only certain types of technologies, as might be the case in an analysis of alternatives that considered operational or other factors.

The important takeaway from the preceding discussion is that materiel developers will use technology to generate the most effective (cost-effective) system approach—that is, the one that yields the best system solution, using technologies and techniques that span EW and cyberwarfare, unless precluded from doing so. The implication is that CERDEC was already using this approach, and it was clear that it expected to use it for future programs.

From there, the discussion with CERDEC I2WD personnel moved to developing a functional view and a technology view to reveal the relationships between the areas we had been discussing.

Figure 5.2 provides a functional view of the relationships and boundaries between and among EW (yellow oval), SIGINT (blue oval), and cyber operations (red oval). Cyber operations are, by far, the most significant (largest) component, which reflects the cyberspace definition provided by the National Military Strategy for Cyberspace Opera-

Figure 5.2
Functional View of Converging Areas According to CERDEC Draft

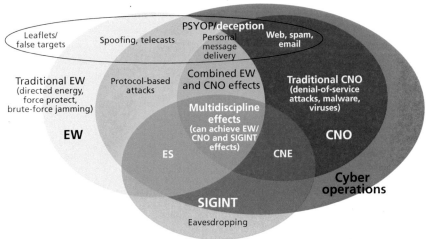

SOURCE: CERDEC I2WD.

tions: "a domain characterized by the use of electronics and the electromagnetic spectrum to store, modify and exchange data via networked systems and associated physical infrastructures" (U.S. Joint Chiefs of Staff, 2006b). Cyber operations overlap approximately 60 percent of EW and 80 percent of SIGINT. It is useful to consider those functions that reside in only one domain: For EW, they are directed-energy and brute force (barrage) jamming; for SIGINT, they are eavesdropping; and for cyber operations, they are a suite of possibilities (e.g., web and email spam, denial-of-service attacks, malware, viruses). MISO (formerly PSYOP, as referred to in the figure) and MILDEC are also shown in the flattened oval at the top of the figure, intersecting both cyber operations and EW, and uniquely containing leaflets and false targets.

While the definitions and terms used in this figure are a bit different and more specialized than those used elsewhere in this monograph, the conclusion that, from a functional view, cyber operations and EW have much in common is supported.

Figure 5.3 provides a technology view of the same universe as Figure 5.2 and also supports the large overlap between cyber operations and EW. Here, cyber is even more dominant than in Figure 5.2, covering almost all of SIGINT and EW. Exceptions are a "planted microphone" that is uniquely in SIGINT and directed energy using advanced electronic steerable array (AESA) antennas. It is likely that ASEA antennas will be used by the Army for cyber effects, too, and will no longer be unique to EW.

All of this points to the blurring of the boundaries in the Army between EW, cyber operations, and SIGINT from a materiel prospective and argues for the consolidation of cyberwarfare and EW.

Using the new lexicon introduced in the TRADOC cyberspace concept plan, this section provided reasons that argue for the consolidation of cyberwarfare and EW in that EW is subsumed—by definition—within cyberwarfare under the current terms of the cyber operations/EM contest CBA. Their technologies and infrastructures are converging. They are operationally converging, as discussed earlier in this chapter and indicated by the argument that if future equipment is designed to provide both cyber and EW capabilities, those opera-

Figure 5.3
Technology View of Converging Areas According to CERDEC Draft

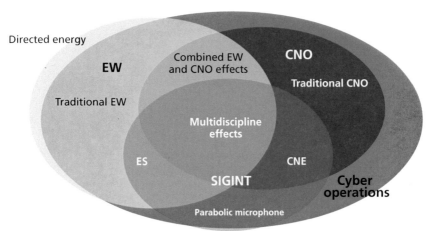

SOURCE: CERDEC I2WD.
RAND *MG1113-5.3*

tions will be conducted with the ability to use in real time whatever capability best satisfies the operational situation. The implications for training are also converging in that the Army must either evolve a set of people with both skills or train sets of people with the constituent skills to work collaboratively in real time.

Electronic Warfare as Fielded Today

Today, the Army is operating EW equipment in Afghanistan to explore effective capabilities. The director of CERDEC's flight activity was quoted in *Aviation Week and Space Technology* as saying, "We demonstrated a quick-reaction Blackhawk capability that could convert UH-60 Black Hawks into an electronic warfare platform within an hour," and in a later part of the article that "we work across all spectrums of ISR, (electronic warfare), cyber (warfare) and information operations" (Fulgham, 2010e). The article also states that

> the basic components of airborne electronic or cyber-attack are
> a sensor that can map an enemy network, the precise location
> of an antenna that feeds the network, and an electronic scanned

antenna that can generate a data stream packed with inquisitive algorithms. That data stream can be beamed into the proper antenna; the target network can be entered and exploited.

Of course, an electronic scanned antenna can also generate the more traditional waveforms associated with EW and directed energy. The intent of the quote is not to suggest the extent to which future capabilities are being used today, but rather the direction in which things are moving.

Chapter Summary and Conclusions

Cyberwarfare, EW elements, and EMSO elements can be consolidated, to an extent. With regard to EW and cyber operations, in particular, technologies, infrastructure, operational use, and personnel training are merging, and these operations will benefit from common doctrines.[3] From public sources, we know that this conclusion is supported by activities and plans in the Army in Afghanistan, the Navy, and the Air Force.

EW and EMSO share some overlapping tasks. In particular, the greatest degree of commonality or overlap exists in the EP component of EW and EMSO tasks.

Additional research and analysis can be undertaken in the context of the Army cyber/electromagnetic contest or an exploration of the two other interconnected dimensions.

Offensive cyber operations (CNA, CNE) and offensive EW (EA, ES) are increasingly similar. In years past, a distinguishing characteristic could have been the lack of information in the signals used for EA. However, new EW concepts involve so-called protocol-based attacks. Thus, it is increasingly difficult to distinguish EA from cyber attack. More importantly, materiel developers are developing tools that do both.

---

[3] Elder (2010) states, "The convergence in effects is real, but there is a danger in believing that there is also a convergence in the expertise to functionally achieve these cyber-electronic-EM effects."

There is an argument that EW and cyber operations should be either combined or assigned to the same staff function to ensure coordination. One of the specific proposals is that many of the technical capabilities most naturally belong under intelligence (e.g., G2/S2). SIGINT includes significant structure (in both DoD and supporting civilian agencies) and is squarely aligned under intelligence. SIGINT has a great deal in common and substantial overlap with cyber operations, EW, and EMSO. It would make sense to align all or part of these capabilities with SIGINT through intelligence staff.[4] There are certainly more courses of action to consider that we have not discussed.[5] It is worth noting that the Navy has already moved to consolidate SIGINT and CNO/cyber operations.

---

[4] According to Elder (2010), "The DoD decision to organizationally place information-related operations under intelligence is a good example of the conundrum associated with the logic currently in use relative to cyber and EW. The challenge now will be to integrate future information-related operations with traditional military operations when the integrator is not clearly the joint force commander." He adds, "Information, cyber, electronics, and spectrum are important to the communications community, the intelligence community, and the operations communities. DoD elected to organize cyber as an intelligence function, but normally, it is the operational commander who is responsible for integration across multiple lines of operation. Clearly, it made good sense in the near term to leverage the technical competencies in the intelligence community relative to cyberspace; now the operational community needs to determine how to integrate these activities with their traditional combat actions to achieve unity of action. Ultimately, there must be only one commander."

[5] As cautioned by Hura (2010), the consolidation of EW and cyber must not compromise the current responsiveness of electronic attack operations, which are not subject to the same legal and authority-control constraints as signals intelligence and CNA/CNE.

CHAPTER SIX

Overlaps Between Public Affairs and Military Information Support Operations

Fundamental areas in the psychological realm, which are closely related but currently kept separate, are PA and MISO.

Comparing Public Affairs and Military Information Support Operations

Public Affairs

As defined in JP 1-02 (U.S. Joint Chiefs of Staff, 2010b), PA includes "those public information, command information, and community relations activities directed toward both the external and internal publics with interest in the Department of Defense." It typically focuses on fact-based truth to maintain credibility. The JP on PA was last revised in May 2005 and covered the following topics:

- the rapid expansion of social media use, even by combatant commands
- Internet access
- unclassified web pages to communicate with external audiences (which could include adversaries and foreign intelligence services).

Military Information Support Operations

According to FM 3-05.30 (HQDA, 2005), the mission of MISO (referred to at the time as PSYOP) is to "influence the behavior of foreign target audiences (TAs) to support U.S. national objectives." Furthermore, it accomplishes this by "conveying selected information and/

or advising on actions that influence the emotions, motives, objective reasoning, and ultimately the behavior of foreign audiences." MISO is currently a core capability under information operations (see, e.g., HQDA, 2003).

Link Between the Two Areas

There has always been a doctrinal link between PA and MISO. JP 3-61 asserts that PA provides information to domestic and international audiences, and it contributes to global influence and deterrence of attacks (U.S. Joint Chiefs of Staff, 2010a). This guidance does not distinguish between adversaries and U.S. audiences. It includes both. Key questions remain, however:

1. Does the definition of PA, and the divide between MISO, need to be revisited?
2. In a full-spectrum, noncontiguous battlefield with nonstate actors, how does one clearly separate and control who gets public information?

A commander would be assuming risk if consideration is not given to the fact that an adversary or even a foreign intelligence service is also a consumer of that information; such parties may even be actively seeking publicly released or available information about military activities. The near-omnipresence of the media, Internet use, and continuous probing on DoD networks highlight the ambiguity and inability to distinguish who the actual consumers of publicly released military information may be. So, why the continual attempt to maintain functional separation between two interrelated capabilities when it is becoming increasingly difficult to distinguish who is actually getting the information?

The Key Concern Regarding the Separation of Public Affairs and Military Information Support Operations

There is a fear that MISO communications could contain less-than-truthful information, which could jeopardize credibility if PA and MISO work together. But ironically, most conventional MISO actions use truthful information, and sometimes the only difference is the recipient of the message, how it is delivered, and to whom it is attributed. Even PA doctrine, e.g., FM 46-1 (HQDA, 1997b), highlights a part of PA that intersects with MISO; one key word seems to be critical to maintaining credibility: consistency.[1]

Key Questions
Additional questions persist:

1. How can the Army gain consistency in messages when the prevailing culture and doctrine say to keep PA and MISO separated but coordinated?
2. How can the Army bring MISO and PA together, given the potentially mischaracterized MISO (as bad and impure)?
3. How can the Army go beyond simply coordinating these efforts and achieve the seamless integration needed for mission success?

When addressing these questions from an information standpoint, we must be mindful of the necessity of operational considerations to do what is necessary to get the mission done. Some uncomfortable decisions must be made, such as whether to make deals with former adversaries during counterinsurgency operations (e.g., as in 2008 in Iraq; see Bruno, 2008).

To ensure consistency, doctrine says that PA and MISO are supposed to coordinate, yet the organizational construct and doctrine calls for physical separation between the two capabilities in staff organizations, which seems to have developed into a cultural divide as well.

---

[1] Credibility, for operational purposes, might be more important for MISO than for PA. PA can make a mistake and explain it. MISO does not have this luxury.

The Problem

There is a problem with separating PA and MISO. It is important to recognize that public affairs officers (PAOs) are not typically part of the Modified Table of Organization and Equipment of brigades, and they may not be at the division, either. This function is assigned as an additional duty to a staff officer when needed. This is relevant for two reasons: The attached PAO has little contact with the commander prior to deployment—and even less staff interaction—so employing the PAO into IO planning is very difficult, and (a related issue) training with the PAO is nearly impossible.

The division between PA and MISO could lead to inconsistent messages, regardless of the intended audience, which can bring credibility into question. Perhaps a firewall of some sort can be employed to obtain the benefits of integrating the two without the downsides of such a move.

Key Challenge: The Firewall Between Inform and Influence

As mentioned earlier, there is apprehension when PA ("inform") and MISO ("influence") are mentioned too close together.

The April 30, 2010, circulating draft revision of FM 3-0 takes the inform and influence activities information task and separates it into two lines of effort: inform and influence. The division is explained thusly:

> Each line of effort has a different purpose and effect. These lines of effort may rely on the same capabilities to accomplish these effects and must be integrated closely to ensure unity of effort in words, images and actions. This is to avoid contradiction or the appearance of contradiction that may undermine the force's efforts. (HQDA, 2010a, para 7-12)

This separation stems from a traditional pejorative separation between "informing" and "influencing," in which informing is held to be value-neutral or -positive and influencing is somehow nefarious and underhanded.

Traditional Missions and Tensions

PA traditionally claims the inform mission and only the inform mission; "inform but not influence" is one of its traditional touchstones (Helmus, Paul, and Glenn, 2007, p. 38). MISO, on the other hand, openly aspires to influence, but exclusively foreign audiences, because it would be both inappropriate and illegal to point the "mind-control lasers" at the American people.

The previous paragraph contains significant fallacies. First, it is naïve to imagine value-free information (see, e.g., Gray, 1989), that it is actually possible to inform without influence, and that as PA operators carry out their inform mission they are not trying to (truthfully) present the Army in the best possible light and maintain support for it (and support for Army recruiting) among a domestic audience. Second, it is equally naïve to say that MISO efforts involve other truly insidious ways to manipulate people or that there is something wrong with trying to influence target audiences, or admitting that you are trying to influence target audiences.

The *real* tension between PA and MISO over inform and influence has to do, first, with the black reputation of MISO as executed by liars and manipulators and, second, with the tension between true and misleading information (discussed next). There may be good reasons to separate inform and influence, but those reasons may disappear if other separations (black efforts from white, for example) are made. Of course, there is a very real concern that must not be diminished. If PA and MISO are to be better integrated, such integration must be done in a way that avoids (and avoids the appearance of) DoD efforts to *inappropriately* influence the American public.[2]

---

[2] The word *inappropriately* is italicized in this text to emphasize the point made in the preceding paragraph—specifically, that all efforts to inform also seek to influence. The difference is whether that informative influence is benign and done in a forthright fashion using only true information or whether it is done using manipulation or falsehood. The former is appropriate for defense information to domestic audiences, and the latter is most definitely inappropriate.

In current MISO doctrine, information is assessed as white, gray, or black based on both its content and its attribution (HQDA, 2003, p. 11-1). For example, a wholly true message attributed to a fictitious author is white content, black attribution. Falsehood in either content or attribution is problematic, because when discovered, it damages credibility. In fact, even the possibility of falsehood damages credibility. While the vast majority of MISO conducted in contemporary operations are completely white, the fact that everyone knows MISO *could* be using falsehoods damages credibility. Similarly, the fact that current doctrine for IO assigns both MISO and MILDEC as core capabilities means that IO can deceive and manipulate, and, thus, such operations remain a source of wariness. If there is black in the toolbox, credibility is lost; in the inform, influence, and persuade arena, credibility is king (Paul, 2008, pp. 38–41).

Black versus white, together with inform versus influence, creates a "firewall" between PA and IO/MISO that "protects" the credibility of PA from the taint of black information and influence. This firewall also inhibits the coordination and integration of PA and MISO themes, messages, and products, increasing the likelihood of information fratricide and decreasing synergies between these two capabilities.

Relationship with Public Diplomacy and Strategic Communication

Defining *strategic communication* is even more contentious than defining IO (Paul, 2009b). There are several possible overlaps between strategic communication and the functional areas described in this monograph. The essence of strategic communication is coordinated efforts to inform, influence, and persuade in pursuit of national policy objectives (Paul, 2009a). If IO (or a large part thereof) is about informing, influencing, and persuading, what separates IO from strategic communication? Perhaps the difference is one of level and component, wherein strategic communication is at the highest levels and involves the whole of government while information operations nest within strategic communication and are conducted exclusively by DoD.[3]

Current views of strategic communication from within the Pentagon assert it is a "process" for integrating and coordinating signals and messages. If that is the case, how does it differ from the current doctrinal integrative role of IO? That is not entirely clear. In fact, some have proposed that IO officers should take on responsibility for coordinating the strategic communication process, and, in some commands, they apparently already do.

If IO constitutes a capability, then such operations nest comfortably under strategic communication. If IO plays an integrative role, perhaps IO integration follows the strategic communication process. If strategic communication is a separate integrating activity, there could be considerable duplication of effort and responsibility between strategic communication and IO. If IO plays an advocacy role, then these operations overlap even more with strategic communication. The relationship between IO and strategic communication depends to a significant extent on how each concept is ultimately defined, but there is considerable possibility of overlap.

---

[3] According to Elder (2010), "It is important to differentiate 'communications strategy' from 'strategic communications.' Strategic communications is the alignment of actions with messages; communications strategy is the process of determining messages and their delivery."

CHAPTER SEVEN

Better Integrating the Technical Realm

Consolidation and coordination need to be considered among the functional areas that relate to IW. There are a number of major factors driving this need:

1. the aforementioned convergence of mediums and applications
2. the expansion of cyberspace in the information environment
3. the need for efficient use of manpower.

In this chapter, we assert that the relevant realms that contain the functional areas pertaining to IW are just two: the psychological and the technical. The psychological, which incorporates the considerations provided for our proposed definition of IO, is focused on content and the target is people. The technical realm is focused on the means to deliver (or prevent delivery of) content, and the targets are machines. Specifically, this chapter considers how the technical realm can best be organized and perhaps consolidated.

Dividing and Conquering the Information Environment: The Psychological and Technical Realms

The psychological and technical realms represent the most compact division and boundary. Based on this assumption, we conclude that subject-matter expertise is best developed by an individual in one or the other but not both at the same time. Table 7.1 represents the realm of the possible for consideration of consolidation.

Table 7.1
Information Warfare: Realms of the Possible

Category	Psychological Realm	Technical Realm (C3IEW)
Functional areas, subareas	MISO, PA, aspects of MILDEC	EA, EP, ES, CNA, CNE, SIGINT, EMSO, IA, operating and maintaining networks (network operations), aspects of MILDEC
Target	People	Machines

Past Consolidation of the Technical Realm

Functional consolidation among the aforementioned technical areas is not new. The intelligence community has long referred to intelligence and electronic warfare (IEW) as a grouping and EW was considered an intelligence function. For acquisition in the Army, the Program Executive Office, Intelligence and Electronic Warfare Systems, serves both areas to an extent.

According to joint doctrine, distinctions among aspects of EW and intelligence are slight. The 2007 version of JP 3-13.1 (U.S. Joint Chiefs of Staff, 2007) notes that "the distinction between intelligence and EW [electronic support] is determined by who tasks or controls the collection assets, what they are tasked to provide, and for what purpose they are tasked."

Years ago, the Army's capstone document for military intelligence doctrine was FM 34-1. There, IEW was defined as something that provides commanders with the ability to visualize the expanded battlespace in command-and-control warfare and to identify where and when they gain information dominance over an adversary (see Peterson, 1997).

The analysis community (U.S. Army, 1997; see also Pawlowski, 1992) uses the expanded term *command, control, communication, intelligence, and electronic warfare* (C3IEW), although it was adopted decades ago (see Peterson, 1997). From our perspective, C3IEW is the technical side of modern IW and is a term that has increased utility. It is now akin to *cyber operations*.

Potential for Electronic Warfare to Be an All-Encompassing Field

EW is defined in doctrine as "military action involving the use of electromagnetic and directed energy to control the electromagnetic spectrum or to attack the enemy" (HQDA, 2009). Spurred by the operational needs to counter RCIEDs, the Army decided to invest in its own EW corps several years ago and put in place a new EW career field (Jordan, 2009).[1] It has been hoped that these operators can fulfill broad responsibilities, such as

- disrupting enemy communication (Vanden Brook, 2007)
- ensuring that U.S and coalition troops can talk to one another (Vanden Brook, 2007)
- preventing the enemy from knowing what friendly forces are doing (Vanden Brook, 2007)
- being the "go-to people for commanders wanting to know how they can exploit the electromagnetic spectrum tactically across their operations" (Jordan, 2009).

In addition to the career field, EW has its own school, training courses,[2] and manpower structure within the Army. Hundreds of additional billets have been allocated to provide EW support across the echelons. This convergence trend suggests that EW operators are fundamental to conducting cyber operations.

---

[1] A 29-series military occupational specialty (MOS) that will include officers, warrant officers, and enlisted personnel.

[2] This training was initially a "tactical course," a three-week session at Fort Huachuca, Arizona, that focused on training soldiers at the battalion level and below; at the brigade level and higher, there was a six-week "operational course," at Fort Sill, Oklahoma. (See Kruzel, 2007.)

First Steps: Deconfliction and Coordination Cells

There is a need for coordination and deconfliction across all functional areas related to IW. With regard to deconfliction of the EMS in particular, COL Laurie Moe Buckhout concluded that "a lack of coordination and networking of EW equipment used in Iraq and Afghanistan is becoming a hazard for U.S. and allied troops" (quoted in Boessenkool, 2009). Expanded IO cells (perhaps called "IW cells") are needed to make decisions. Existing IO cells are vital but should expand and transform into what we call IW cells, replete with specialists in both the psychological and technical realms, e.g., EW, MISO, CNO, and network operations. These personnel will need to be able to plan, operate, and integrate all IW-related activities.

Such cells will be necessary for supporting the commander in making the proper trade-offs and evaluations of

- cyberspace gain or loss
- information gain or loss
- EMS gain or loss.

Like today's IO cells, IW cells are needed at all echelons.

Some Sharing of Manpower for Efficiency

For the sake of efficiency, cross-trained specialization is needed. However, there is a limit to this cross-training. Areas that fall within the psychological realm need focused expertise. The same is true for the technical realm. In other words, those trained in the technical realm (e.g., to deliver content) cannot be expected to cross-train into the psychological realm (e.g., to develop information content), and visa versa.

There are limits to consolidation in any of the individual realms related to tasking of personnel. For example, not all the tasks (or even most) in the technical realm can, or should, be consolidated.

Consider the tasks associated with three areas: IA, network operation and maintenance, and CND. The same person or position may not be ideal for carrying out all three.

First, lessons learned from the network operations community (see Porche et al., 2010) is that dual-hatted IA personnel were left little time to focus on security demands in addition to their primary tasks, like maintaining connectivity of the network. Note that businesses and organizations with IT staff experience the same personnel issues: System administrators have little time (and sometimes little motivation) to function dually as network defenders.

Second, IA functions (such as achieving certifications for equipment and updating software patches) support but are not equivalent to CND. CND is fundamentally different. It is about countering adversaries who have penetrated Army networks or are in the midst of trying. Such a fight—which may occur in real time—will take place in Army networks, and the personnel involved need the appropriate focus, training, clearance levels, and experiences for such operations.[3]

Furthermore, the large signal corps currently dedicated to operating and maintaining LandWarNet needs to remain single-purposed. Further analysis will be necessary to identify ideal command-and-control relationships for various technical capabilities at different echelons and the specific technical tasks that will need to be performed at those different echelons.

A Cadre of Cyber-Electronic/Electromagnetic Warriors

The Army eventually needs to either create a new "cyber-electronic" or "cyber-electromagnetic" career management field or transform an existing one to support all the technical realms of IW. This would serve as a first step toward a new branch for cyber-electronic warriors that can be utilized to cover the C3IEW areas described here. This grouping includes EW and spectrum managers and falls within the technical

---

[3] Elder (2010) notes, "While cyber security can be assigned to "security forces," cyber defense requires the full participation of all communities, under the leadership of the operational commander."

realm of IW.[4] A similar argument can be made for the psychological realm we describe in this paper.

In terms of expanding an existing field, EW should be a prime candidate. Spurred by the operational needs to counter RCIEDs, the Army has already invested in a new EW career field with hundreds of billets and a training pipeline.

The importance of sustained career paths in these areas cannot be overstated. This would allow personnel to receive repetitive assignments to hone proficiency and would help attract the best. In particular, a cadre of cyber-electronic specialists, i.e., strategists, is needed to continuously develop new ways to apply cyber-electronic power and new tactics, techniques, and procedures at all echelons.

Conclusions on Redefined Boundaries

Information warfare as we define it has two aspects: the psychological and the technical. Information content is key for the psychological part; the means to deliver content (or counter its delivery) are key to the technical part. Ultimately, it makes sense for most of what falls into the psychological realm (shown in Table 7.1) to be redefined as IO and for most of what falls into the technical realm to be considered cyber-electronic, with the exception of operating and maintaining the network. In the next chapter, we discuss the details, options, and other considerations with regard to better organizing the psychological realm for IO.

---

[4]   In the case of EW, the Army recently created a new career management field that provides for a new MOS for officers, warrant officers, and enlisted personnel. Hundreds of billets (more than 3,000) have been created, although not yet all are filled. The specific career management field identifiers for electronic warfare are to be Functional Area 29 for officers, MOS 290A for warrant officers, and MOS 29E for enlisted personnel.

The signal corps has the 25E enlisted specialty for spectrum management. Previously, noncommissioned officer spectrum managers were tracked with a skill identifier attached to a preexisting MOS. The skill identifier for enlisted personnel (for spectrum managers) was not found to be satisfactory because these spectrum managers were often retasked outside the spectrum specialty.

CHAPTER EIGHT

Better Integrating the Psychological Realm

In Chapter Four, we proposed a new definition for IO that is focused on what we call the psychological realm, or content-based capabilities. In this chapter, we recognize that organization and staffing changes should be commensurate, but there are significant considerations that must occur before courses of action can be clarified. This chapter identifies these considerations for the psychological realm.

Importance of IO Staffing and Available Options

Much of the IO reform debate centers on how, exactly, IO and various core, supporting, and related capabilities ought to be assigned to staffs. A clear vision of the nature and goals of IO should significantly advance the debate.

In the most current joint structure, the IO chief is the J-39, a subordinate staff under operations (J-3). If IO is to be something other than what it is now in joint doctrine, where should it be, and where should the bits that currently constitute it be located? Because of the targeting and effects components of inform and influence, we maintain that IO should fall under fires at the division level and above and should be incorporated with fires at the brigade level and below.

Proposals That Have Been Considered

In the psychological realm (e.g., content-based capabilities, such as MISO), proposals have located the capabilities in several different places.[1] Some have proposed aligning content-based capabilities under civil affairs. This is reasonable if such capabilities are focused only on information engagement and if information engagement is focused only on civilian populations, but it falls down if influence supports operations beyond civil affairs or includes targets beyond civilian populations. Others, including the 2008 version of FM 3-0 have proposed new "information engagement" staff (as 7, which is training under the Napoleonic staff tradition). Still others consider influence to be a form of fires (albeit nonlethal fires) or effects, which would align these efforts under operations.

Many of the other proposals make sense, depending on how they resolve some of the other tensions discussed here. There are several additional issues to be aware of when considering staffing relationships, as we discuss next.

2008 FM 3-13 Proposal. The 2008 version of FM 3-0 virtually dismantles the traditional IO concept, reassigning the composite capabilities back to logical staffs. The proposed details are as follows: MILDEC goes to the Army G-5; OPSEC goes to G-3, and the counterintelligence portion of OPSEC goes to G-2. The EA portion of EW and the CNA portion of cyber operations both go to G-3 (fires), while EWS and CNE go to G-2. EP and CND join IA and are assigned to G-6. PA remains assigned to PA, outside the staff hierarchy, but is coordinated with civil-military operations (assigned to G-9) and MISO, combat camera, and defense support to public diplomacy (all assigned to G-7, information engagement) through the information engagement working group.

In our view, establishing the G-7 as the information engagement staff and assigning to it influence messaging capabilities will only serve to marginalize the influence component of IO. After the commander and the chief of staff, the J/G-3 has much control. If one wants something (like influence) to be treated as though it is important and has priority, putting it outside or below J/G-3 is *not* the way to accom-

---

[1]   Elder (2010) notes that, "before designing organizations, it is useful to determine what functions they will perform. This has yet to be performed in the psychological (cognitive) realm."

plish that. Further, separating information engagement from operations excuses operations from the need to integrate with engagement and engagement from the need to integrate with operations, making the marginalization complete. Only a commander firmly committed to maximizing influence information effects would be able to effectively integrate influence with maneuver and fires under this staffing scheme, and then only through constant vigilance. We are not alone in our concern regarding the G-7 concept.[2]

A Staff Structure for Information Advocacy

If IO is going to be about advocacy for consideration of effects in the cognitive and information domains, the staff structure needs to reflect that. On one hand, having a whole staff section committed to an information activity (such as the G-7) gives information some prominence on the staff. On the other hand, unless the G-7 (or a representative) is included in a substantial number of routine working groups in other staff areas (notably G-3 and G-5), information advocacy in those other staff functions will remain difficult.[3]

As we noted earlier, the preeminent position in a staff after the commander, deputy commander, executive officer, and chief of staff is the G-3. This suggests two possible locations for the information advocate: first, in the G-3, with a seat on the fire coordination support cell, a liaison role with G-5, and regular personal access to the G-3; second, as a special adviser to the commander outside and above the normal staff hierarchy, in very much the same kind of relationship now

---

[2] For example, Rosin (2009, p. 7) notes,

> Under the new paradigm, the extant purpose of the G-7 is to integrate the PA and PSYOP functions and be the (Public Affairs Officer) traditional function and impinge on what the PAO perceives to be his role on the staff. Moreover, the PAO is a special staff member for the commander and has direct access to him while the G-7 works under the supervision of the Chief of Staff (CoS) and does not have direct access. As a result of this overlap of functions and roles, friction in the staff will invariably occur . . . one is left to wonder how much value this really provides the commander.

[3] This is important for orders production and the targeting meeting.

enjoyed by the PAO. In fact, if inform and influence are combined and become the core emphasis of the information advocate, that advocate (with proper training) could wholly replace the role of the PAO. The PAO is a very minor staff member. PAOs generally do little staff planning and operate independently to extend command information and work with the external press. This is particularly true for staffs that do not have an organic PAO (which is to say, most staffs). Alternatively, the PAO (again, with proper training) could take on the role of the information advocate.

An "information advocate" could equally be considered an "information warrior" or "cognitive effects proponent" or some combination thereof.

Other Considerations

Separating Lethal and Nonlethal Capabilities

All the capabilities in the psychological realm, however constructed, are unambiguously nonlethal (nonkinetic). However, lethal (kinetic) capabilities create effects in the information domain. Fires and maneuver both communicate and contribute to (or detract from) efforts to inform, influence, and persuade; the destruction of computers, networks, emitters, and receivers affects technical IO capabilities. Nonlethal effects (or fires) must be coordinated with, and in some cases depend on, lethal fires in order to function properly.

This dependence only works one way. That is, lethal fires are in no way dependent on nonlethal fires to realize destructive effects.[4] This poses a problem when and if nonlethal fires (or effects) are coordinated by a separate cell from lethal fires (or effects), as is often the case in current operations and is recommended in FM 3-0 (both current and proposed). What happens when one organization's operations are dependent on integration with the operations of another organization,

---

[4] Lethal fires may be dependent on MISO/civil affairs for population control to be able to conduct fires in some circumstances. It depends on how *nonlethal fires* is defined.

but that second organization has no dependence on the first? Coordination becomes a much lower priority for that second organization.[5]

Given that the military already has a significant bias toward kinetic, lethal, or other "traditional" military undertakings, separating lethal and nonlethal effects further deprioritizes the latter, effectively putting their accomplishment at risk.[6]

If information effects are to be a priority, staffs should be structured and filled to reflect that. A 2007 *Military Review* article states that

> since IO and PA are as important on the modern battlefield as Congressional Affairs is on the home front, it might be time to consistently assign some of the best and most qualified officers to these positions. Perhaps the top two officers in a battalion, brigade, or division should be PA and IO officers. (Chiarelli and Smith, 2007, p. 12)

Designing an Organization to Be Stovepiped Versus Centralized

Virtually all core, supporting, and related capabilities of IO are inclined to operate inside their own functional bounds (often pejoratively referred to as "stovepipes" or "cylinders of excellence"). This tendency is no doubt part of the driving imperative for IO integration in the first place. When does it make sense to retain these capabilities as separate organizational entities, and when does it make sense to combine and centralize them?

On the technical side, cyber operations, EW, EMSO, and so on are all separate organizations. Similarly, on the content side, PA and MISO have many overlapping objectives, capabilities, and processes but are separate (and sometimes competing) organizations.

---

[5]   As suggested by Elder (2010), "Given that we are operating in a networked world, an alternative to a tightly controlling "single integrating authority" is to establish a unity of action process using a Web 2.0 approach but with a more formal governance mechanism that is ultimately responsible to the commander. This approach could be useful in other areas (such as those where non-military capabilities are employed for unified action), but few leaders have experience with this form of enable-influence control model."

[6]   And disrupts the efforts or effects of both.

If the vision of IO includes certain capabilities working together in seamless harmony, then those capabilities need to be combined or placed under the tight control of a single integrating authority.

Personnel Options

Indirectly related to the different ways to staff information-related capabilities is the role and future for Functional Area 30, IO officers.

> The current FA30 (IO) program attracts officers from across the basic branches; however, most have little experience in core IO elements such as PSYOP, computer network operations (CNO), electronic warfare (EW), military deception, and operational security (OPSEC). (Brown, 2005, p. 39)

Thus, an individual FA-30 is a "'Jack or Jill' of all IO 'trades'" (Wass De Czege, 2008, p. 18).

The future role of FA-30 personnel depends largely on the desired vision for IO, and the answers reached for the various questions posed throughout this monograph. If IO continues to be about integration (as in current doctrine), then an FA-30 should handle that integration. If IO is to be about information advocacy, then an FA30 should sit in that advocacy position. If IO is an operational capability, then FA30s need either a lot more training or to get out of the way.

Current FA30 training prepares an IO officer to be an integrator. If the role will be something other than or more than that, changes need to be made. Changes could include more training, a different career path, or different prior functional area experience requirements for staff who join FA30, such as experience in MISO or PA (currently, there are no specific requirements; Brown, 2005, p. 39).

The future of FA30 has been discussed elsewhere. Recommendations for that future include

- "merging the IO functional area (FA) and the Psychological Operations (PSYOP) branch into one specialty under the umbrella term 'information operations'" (Rohm, 2008, p. 108)[7]
- eliminating FA30 and allowing a MISO officer to "assume most, if not all, of the so-called IO coordinating functions" (Boyd, 2007, p. 70)
- letting FA30 become the information engagement officer, while letting FA29 (EW) become the cyber-electronic warfare officer, thus bifurcating the existing IO concept (Whisenhunt, 2009, p. 29)
- "working to update DA Pam 600-3, Chapter 20 to reflect the title of FA30's to read 'Information Engagement Officer,'" which the U.S. Army Information Proponent Office (2009b, p. 8) announced it was doing in April 2009.

Command Considerations

How Much Is the Commander's Responsibility?

While contemplating changes necessary to realize a new vision of IO, another issue that requires consideration is the extent to which IO will be the commander's responsibility.

Ultimately, everything is the commander's responsibility, but perhaps the main reason for the existence of the staff is to subordinate some of that responsibility so that the commander does not have to do everything alone. A worrying number of proposals and concepts in the information arena, however, heap more responsibility directly at the feet of the commander.

For example, as a *Military Review* article suggests,

> Although IO and PA officers, effects coordinators, and others provide critical staff support to the information campaign, commanders must take the lead and be intimately involved in ensuring that the information aspects of military operations are con-

---

[7] As noted elsewhere, PSYOP is now referred to as MISO.

sidered in every action we undertake. It is that important to our success. (Chiarelli and Smith, 2007, p. 10)

Similarly, discussions at the 2008 U.S. Army Combined Arms Center Information and Cyberspace Symposium included the view that "commanders own cognitive effects in a direct and personal way because such effects are the cornerstone of their battle command" (U.S. Army Combined Arms Center, 2008). This reasoning supported the disaggregation of IO capabilities back to related staffs (as proposed in FM 3-0) so that they were all available to be called upon by the commander.

The new draft of FM 3-0 asserts, "When conducting inform and influence activities, commanders must determine how these activities affect the perceptions and actions of audiences" (HQDA, 2010). If we understand correctly, that is another way of saying "it is the commander's responsibility."

Just how much responsibility for IO do we want to put directly on the commander's shoulders? While direct responsibility for IO would ensure that if such operations are not a priority it would be the commander's fault, we wonder how well prepared most commanders are for that role. As noted previously, IO lies outside the traditional core of military culture and thinking.

Proposals to Balance a Commander's Responsibility

We are aware of two proposals that might balance a commander's acceptance of direct responsibility for IO. The first is the presence of an information advocate, a notion that has been discussed repeatedly throughout this monograph. The second involves a change in the requirements for producing the commander's intent. Professor Dennis Murphy at the Army War College has proposed that guidance for commander's intent be changed to require the specification of a commander's intended cognitive or information end state (Murphy, 2008). A somewhat banal example is that the traditional commander's intent might be to "remove the insurgent threat from village X." Subordinates executing based on this guidance have the whole military toolbox open to them: They could level the village, cordon and search, or use a variety of softer approaches.

Now imagine the implications if the following information end state is specified: "If possible, leave the population of village X neutral to our presence." That significantly changes the approaches that subordinates are likely to take. It also allows the commander to assign explicit priorities to kinetic versus informational outcomes, or short-term versus long-term outcomes. There may be cases in which the informational end state does not matter, and the commander's intent really is strictly kinetic. In the vast majority of situations, however, that will not be the case. If commanders are forced to think about and be explicit about communication and information end states, their subordinates will be similarly forced.[8] Under this construction, while the commander accepts responsibility for conceiving the information end state, subordinates naturally accept more responsibility for achieving it.

Resolving Overlaps Between Public Affairs and Military Information Support Operations

In an earlier chapter, we described the problem of the separation of PA and MISO and how this could lead to poor coordination. We explained the existing cultural divide between the two and the prevailing feeling that the traditional MISO mission or approach could hinder (or taint) PA. We acknowledge the importance of better integration while protecting against the appearance that the Army might be trying to inappropriately influence the American people. In this section, we provide possible courses of action for resolving barriers to MISO and PA integration.

---

[8] Elder (2010) says, "One reason that acceptance of responsibility for information operations may pose a problem for some commanders is that end-states in commander's intent statements often are described in military rather than PMESII terms. For example, a typical end-state may be expressed as "insurgents removed" when the actual end-state is "population feels secure. A more accurate and complete end-state should stimulate consideration of information environment activities."

Option 1: Maintain the Status Quo (Do Nothing)

The first course of action is to leave PA and MISO as is, where inform and influence are separated and deconflicted but often not truly being integrated.

Option 2: Combine the Two Areas

The integration of inform and influence, PA and MISO, can be done without adjustment. An information engagement cell[9] would be constructed to combine the capabilities that were a part of IO and deal with the human dimension (i.e., MISO, PA, military support to public diplomacy, and combat camera). This could help invoke cross-coordination and integration, minimizing inconsistency in information that is distributed across audiences. Those audiences could include U.S. allies, local populace, adversaries in direct contact, and potential adversaries that are watching U.S. actions as well.

The argument for avoiding a separation and ignoring the concerns is as follows: The stigma of having a function, cell, or people who may use information differently for each target should not be a reason to create seams in the Army organization if the ultimate goal is mission accomplishment. Unless there is a prevailing case against it, the question remains: Why not capitalize on the potential unity of effort that could result?

Option 3: Combine the Two Areas but Abandon the Black

A third option is to integrate inform and influence, and PA and MISO, with some constraints. This course of action would require that MISO (formerly PSYOP) abandon its black toolbox and use only truthful information and attribution.[10] (This is not much of a change; the vast majority of MISO efforts in contemporary operations are wholly

---

[9] Information engagement is the integrated employment of PA to inform U.S. and friendly audiences; MISO, combat camera, U.S. government strategic communication and defense support to public diplomacy, and other means necessary to influence foreign audiences; and leader and soldier engagement to support both efforts.

[10] As suggested by Elder (2010), "A way to 'abandon the black' in MISO (PSYOP) without abandoning this tool completely is to doctrinally place the 'black' side of PSYOP into MILDEC."

truthful.) The firewall would then be between MISO and residual black PSYOP, still dividing white from black, but now with PA and "white hate" MISO on the same side. The relocated firewall would now separate virtuous persuasion from manipulation and falsehood while retaining both forms of capability. The two sides still require coordination, deconfliction, and integration.

Based on the results of our study, option 3 is the progressive course of action that we tentatively recommend. More analysis and study are needed to produce firmer conclusions regarding MISO and PA consolidation. Specifically, the aforementioned concern about creating the perception that DoD or the Army might be seeking to inappropriately influence the American people under such a construct must be addressed.

CHAPTER NINE

Conclusions and Recommendations

Today, the wireless and wired mediums are converging. Computer and telecommunication networks are becoming one and the same. Transmission of digitized packets over Internet-protocol networks is rapidly supplanting the old technology (e.g., dedicated analog channels), particularly for information sharing and media broadcasting. In short, the information environment is changing in fundamental ways, and Army doctrine needs to change to accommodate those changes. The Army is aware of many of the challenges in this area, and it is in the process of pursuing change. As noted earlier in this monograph, IO is a "moving target," and several changes had been implemented or were under way at the time of its publication in 2012. Important questions and important decisions remain, however.

This monograph identified the implications of these trends and reconsidered the resulting boundaries of Army cyber operations in particular and information warfare more broadly.

Authorities for Cyber and EW Are Currently Not the Same

Today, there is a difference in the existence, understanding, and use of authorities between offensive EW (e.g., ES and EA) and offensive CNO (e.g., CNE and CNA). The use of EW capabilities at all echelons has precedent. For cyber to enjoy the same level of permission, understanding, and use, many policy changes are needed. In addition, training will have to be commensurate. Other differences exist with respect to the classification levels and authorities currently used for certain EW operations compared to CNO.

This is understood in the Army (see TRADOC Combined Arms Center, 2009) and will require more time. Therefore, we caveat our recommendations by stating that we assume that the proposed doctrinal changes we recommend will be implementable. It is worth noting that a capability can be organic to a tactical unit even if the authority to use it rests at the highest echelons. An example is a tactical nuclear weapon.

Challenges and Practical Considerations

This monograph would not be complete without discussing the practicality of implementing now some of the doctrinal changes proposed into operational capabilities. Throughout, we discussed some difficulties with operationalizing the doctrinal changes recommended. We summarize them in Table 9.1. Still, the "RF aperture" will increasingly be an entry point for CNA/CNE.

As a result, we conclude that EW and CNO could and should eventually share the same people, process, and technology to avoid duplication of effort or working at cross-purposes, at least at some point in the near future. We understand that the Army has already instituted some changes moving in this direction The same can be said for EMSO, as well as certain aspects of intelligence operations.

Doctrinal Reorganization

Figure 9.1 illustrates our conclusions with respect to doctrinal reorganizations. Under this taxonomy, IW would have two subdivisions: information technology operations and information and influence operations. This organization explicitly recognizes the fundamental difference in targets and methods. The old organization of CNO and EW would fall under information technology operations, organizationally formalizing the convergence of the two disciplines. We would add EMSO to information technology operations. We place both MISO and PA under inform and influence operations, recognizing the close relationship between them and the need for close coordination. MILDEC resides under inform and influence operations but is fire-

Table 9.1
Challenges, Practical Considerations, and Potential Resolutions

Challenges	Practical Considerations	Potential Resolutions
Authorities for cyber and EW are currently not the same	This tension is not likely to be resolved immediately (short term); new legal authorities for the use of certain cyber operations at the operational and tactical levels remain debatable for now.	Authorities' issues have been flagged at high levels in the Army, DoD, and the U.S. government. They can be resolved in the future, enabling EW and cyber to move closer from an operational standpoint.
Existence of black vs. white information barriers among influence operations	Arguments remain for a need for some amount of black information capabilities.	Black information can be exclusively MILDEC.
Personnel constraints	Currently, a zero-sum constraint with regard to billets for all areas related to IW has to be assumed.	Transfers and retraining will support personnel needs.
Speed and process for change for joint doctrine and Army IO doctrine	Joint doctrine in IO is being reconsidered and potentially being cemented. Joint doctrine requires a broader consensus and is not likely to change again for some time.	The projected joint IO definition is expected to be finalized soon, and drafts suggest that the definition is broad enough to support further refinement in Army doctrine in the manner we recommend.
What to do with FA30	Existing FA30 (IO) officers are used in the field today based on training dictated by existing (but old) doctrine.	Current IO officers can be retrained, and retasking future ones can use modified FA30 training requirements.

walled off from other capabilities and included only to ensure decon-fliction and coordination.

Information Warfare Cells Are Needed at the Strategic, Operational, and Brigade-and-Below Levels

Existing IO cells are vital but should transform into what we call IW cells, staffed with specialists in both the psychological and technical realms (e.g., EW, MISO, CNO specialists). With regard to

Figure 9.1
Suggested IO Reorganization

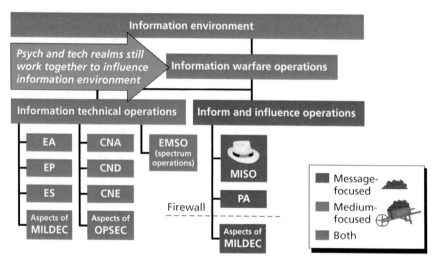

IW activities, these personnel will need to be able to plan, operate, and integrate all IW-related activities and help direct the actions of operators—MISO teams, civil affairs teams, associated EW assets, and so on.

Such cells will be necessary for supporting the commander in making the proper trade-offs and evaluations of the following issues:

- cyberspace gain or loss
- information gain or loss
- EMS gain or loss.

Like today's IO cells, IW cells are needed at all echelons.

For operational purposes, EW is planned to be a part of the fires cell as a fires warfighting function. This is just one course of action and another argument with historical precedent: EW has been supported under the intelligence warfighting functions or cells in the past.

Reviews of how well this arrangement worked are, at best, mixed. This monograph is agnostic on the subject of such courses of action.[1]

Some Cross-Training Is Needed, but Technology and Content Areas Require Specific Expertise

For the sake of efficiency, cross-training is needed, i.e., personnel with more than one specialty. But there is a limit to this cross-training. Those areas that fall within the psychological realm need focused expertise (see Table 9.2). The same is true for the technical realm. In other words, those trained in the technical realm (e.g., to deliver content) cannot be expected to cross-train into the psychological realm (e.g., to develop

Table 9.2
Information Warfare: Realms of the Possible, Renamed

Category	Psychological Realm	Technical Realm
Functional areas, subareas	MISO, PA, aspects of MILDEC	EA, EP, ES, CNA, CNE, SIGINT, EMSO, IA, operating and maintaining networks (network operations), aspects of MILDEC, aspects of OPSEC
Target	People	Machines
Realms renamed	Inform and influence operations (IIO)	Information technical operations (ITO) or cyber-electronic operations

---

[1] Elder (2010) explains that

> the communications community has traditionally been the technical provider of information services, which were then used by the intelligence and operations communities. Now that the intelligence community has begun to play a major role in the technical aspect of the information environment, this has led to organizational changes to leverage these competencies. One of the questions for the Army (and other services) is the following: Does technical expertise lead to operational expertise? Ultimately, the Army (and others) must decide on the best way to integrate operations in (and through) the information environment with traditional military ops. This is a process question which should inform organizational decisions; however, in our hierarchical society, it is easier to organize first and then challenge the new organization's leader to develop the processes; however, the processes are then typically confined to the resources available to the organization rather than those available in the enterprise.

information content), and vice versa. Along these lines, it is vital to avoid the misuse of personnel (i.e., people tasked to do things they are not trained to do).

There are limits to consolidation in any of the individual realms. As documented by Porche et al. (2010) lessons learned from the network operations community show that dual-hatted IA personnel are left little time to focus on security demands compared with their primary tasks, like maintaining connectivity of the network. Similarly, businesses and organizations with IT staff experience the same personnel issues: System administrators have little time (and sometimes little motivation) to function dually as network defenders. In short, not all of the tasks in the technical realm can be consolidated. Furthermore, the large signal corps currently dedicated to operating and maintaining LandWarNet needs to remain single-purposed.

Doctrine Needs to Be Revised

The existing doctrine for IO (FM 3-13) has to be revised. Because there are genuine disputes regarding both the terminology and the concepts of IO, simple clarification will not resolve them. Decisions must be made. We know that the Army is again attempting to revise FM 3-13 (after a similar such attempt in 2009). Should the Army succeed in reaching consensus and advancing this new doctrine, many of the questions (posed genuinely as questions) in this monograph may be more definitively resolved.

OPSEC and MILDEC: What to Do

Unlike the other traditional pillars of IO, MILDEC and OPSEC are both "phantom" capabilities. That is, there is no force structure dedicated to MILDEC or OPSEC. They could be moved out of IO doctrine. Justification for such a change is as follows.

Regarding MILDEC, certain aspects of EW and CNO do include MILDEC, but it is exclusively tactical (and technical) MILDEC.

When MILDEC is operational or strategic, it is planned and staffed at the commander's discretion. Significant operational deception requires integration well beyond the traditional IO domain and certainly includes maneuver elements; it has to be under mission command. EW and CNO could certainly support such broader MILDEC efforts, but no more or less than any other capability in the commander's toolbox.

OPSEC is also a capability without structure. OPSEC should be everyone's responsibility. It has suffered because it has been nested under IO, which makes it easy to ignore.

The Army Needs to Develop a Cadre of Cyber-Electronic or Cyber-Electromagnetic Warriors with a Dedicated Career Path

The Army eventually needs to either create a new cyber-electronic or cyber-electromagnetic[2] career management field or transform an existing one to support all the technical realms of IW. This would serve as a first step toward a new branch for cyber-electronic warriors that can cover the technical functional areas we describe. This grouping includes EW and spectrum managers and falls within the technical realm of IW.[3] A similar argument can be made for the psychological realm.

In terms of expanding an existing field, EW should be a prime candidate. Spurred by the operational needs to counter RCIEDs, the

---

[2] This could be called "spectrum-electronic."

[3] The signal corps' MOS 25E enlisted specialty for spectrum management was created a number of years ago. Prior to the creation of this specialty, noncommissioned officer spectrum managers were tracked only with a skill identifier attached to a preexisting MOS. The skill identifier for enlisted personnel (for spectrum managers) was not found to be satisfactory because these spectrum managers were often retasked outside of the spectrum specialty. There is a skill identifier for commissioned officers, but it is dormant.

In the case of EW, the Army recently created a new career management field that provides a new MOS for officers, warrant officers, and enlisted personnel. Hundreds of billets (greater than 3,000 personnel) have been created, although not all have been filled. The specific career management field identifiers for EW are to be FA29 for officers, MOS 290A for warrant officers, and MOS 29E for enlisted personnel.

Army invested in its own EW corps (Jordan, 2009) several years ago by putting in place a new EW career field.[4] Hundreds of billets have already been created, and the training pipeline has begun to be filled. It is hoped that these operators can fulfill broad responsibilities, including some of the following areas in the technical realm:

- disrupting enemy communication (Vanden Brook, 2007)
- ensuring U.S and coalition troops can talk to one another (Vanden Brook, 2007)
- preventing the enemy from knowing what friendly forces are doing (Vanden Brook, 2007)
- Being the "go-to people for commanders wanting to know how they can exploit the electromagnetic spectrum tactically across their operations" (Jordan, 2009).

The importance of sustained career paths in these areas, like the ones developed for EW, cannot be overstated. This would allow personnel to receive repetitive assignments to hone proficiency. In particular, a cadre of cyber-electronic specialists, i.e., strategists, is needed to continuously develop new ways to apply cyber-electronic power and new tactics, techniques, and procedures at all echelons.

From a personnel standpoint, the main objective is a career pathway enabling a balanced force of intelligence, technicians/signal, and operations to ensure that diverse, flexible, and operationally feasible solutions are available for implementation without seams. An example of a long-term plan needed for a "sustained path" is found in the Navy, as documented in Appendix F of this monograph.

---

[4]   A 29-series MOS that will include officers, warrant officers, and enlisted personnel.

Future Work

Considering Whether Network Operations Should Be Organizationally Separate

A key question remains: Where do traditional network operations (as defined in FM 6-02.70) best fit in the technical realm? We argue that in a cyber-electromagnetic contest, network operations fit everywhere and nowhere. Portions fit in the technical realm—to an extent—and portions fall into the psychological realm. But overall, network operations are probably best considered a separate entity. Simply put: Networks are the superstructure for cyber operations. The Army depends on its networks for its own communication and situational awareness needs. From a practical standpoint, such efforts are too substantial to ever be enveloped by any other construct. Network operations require an order of magnitude more personnel than all other technical areas. Concepts of such operations tend to focus more on the user than on ambiguous threats, which is both a curse (see Porche et al., 2010) and a blessing. In summary, the planning, engineering, installation, operation, and maintenance of a network at the enterprise and tactical levels is a substantial, continuous effort: It is an area unto itself.[5] This is something that requires more study and was beyond the scope of our research.

Considering Capability Integration from a Process Standpoint

As suggested to us by our reviewers (especially Elder, 2010), it would be useful to explore the integration of the capabilities in the functional areas from a process standpoint rather than an organizational one. The purpose would be to examine the synergistic impact of cyber-electronic operations unencumbered by traditional, hierarchical structures.

---

[5] Elder (2010) notes that "networks are also the foundation for psychological (social, cognitive) interactions as well as the physical connections of machines and people. However, the premise that NETOPS 'is an area unto itself' is an important point. It is a technical specialty historically associated with the communications community, and while integration with intelligence and operations is critical, it is neither intelligence or operations."

Existing Terminology, Doctrine, and Ongoing Studies

Electronic Warfare

JP 3-13.1 defines EW "as any action involving the use of electromagnetic (EM) or directed energy (DE) to control the electromagnetic spectrum (EMS) or to attack the enemy" (U.S. Joint Chiefs of Staff, 2007). According to JP 3-13.1, the three components of EW are EA, EP, and electronic warfare support. The EW CBA, issued by the TRADOC Analysis Center, added a fourth component: EWI.

Spurred by the operational needs to counter RCIEDs, the Army decided to invest in its own EW corps several years ago by putting in place a new EW career field (Jordan, 2009).[1] It is hoped that these operators can fulfill broad responsibilities, including

- disrupting enemy communication (Vanden Brook, 2007)
- ensuring that U.S and coalition troops can talk to one another (Vanden Brook, 2007)
- preventing the enemy from knowing what friendly forces are doing (Vanden Brook, 2007)
- being the "go-to people for commanders wanting to know how they can exploit the electromagnetic spectrum tactically across their operations" (Jordan, 2009).

---

[1] A 29-series MOS that will include officers, warrant officers, and enlisted personnel.

In addition to the career field, EW has its own school, training, and manpower structure in the Army.[2] Hundreds of additional billets have been allocated to provide EW support across the echelons. The convergence trend described here suggests that EW operators are fundamental to conducting cyber operations.

EA uses electromagnetic, directed-energy, and antiradiation weapons to attack personnel, facilities, or equipment with the intent to degrade, neutralize, or destroy enemy capability. EA is considered a form of fires and includes offensive and defensive countermeasures. Examples of offensive EA includes jamming an adversary's command-and-control system, using antiradiation missiles to suppress air defense, using directed energy to disable adversary's equipment or capabilities, and employing electronic deception techniques to confuse an enemy's intelligence, surveillance, and reconnaissance system. Defensive EA actions include the use of flares, jammers, towed decoys, counter-RCIED systems, and other assets for self- and force protection (U.S. Joint Chiefs of Staff, 2007).

ES or EP activities seek to protect personnel, facilities, and equipment from the effects of friendly or enemy use of the EMS that degrade, neutralize, or destroy friendly combat capability. Examples of EP activities include spectrum management, EM hardening, emission control, and use of wartime reserve modes. Electronic protection actions attempt to ensure friendly use of the EMS through tactics as frequency agility in a radio or variable pulse repetition frequency in radar (U.S. Joint Chiefs of Staff, 2007).

EWS is actions tasked by operational commanders to search for, intercept, identify, and locate sources of intentional and unintentional radiated electromagnetic energy in order to recognize threats and to target, plan, and conduct operations. ES data can be used to produce SIGINT, provide targeting for electronic or destructive attack, and produce measurement and signature intelligence (U.S. Joint Chiefs of Staff, 2007).

---

[2] This training was initially a "tactical course" that was a three-week session at Fort Huachuca, Arizona. It focused on training soldiers at the battalion level and below. At the brigade level and higher, a six-week "Operational Course" existed at the Fort Sill, Oklahoma. (See Kruzel, 2007.)

Table A.1 shows the details of the aforementioned divisions as outlined in Army doctrine (HQDA, 2009).

The U.S. Army EW Integrated Concept Development Team first introduced the nondoctrinal term *EW integration*. EWI refers to capabilities that synchronize and coordinate EW into the commander's overall campaign or operations plan. The vision is for EW operations to support all key operational ideas across the full spectrum of war. The key operational ideas are shaping and entry operations, decisive maneuver, operational maneuver from strategic distances, concurrent and subsequent stability operations, network battle command, distributed support and sustainment, and intratheater operational maneuver (TRADOC, 2007).

Electromagnetic Spectrum Operations

EMSO involves planning, operating, and coordinating the use of EMS. The objective of EMSO is to enable spectrum-dependent devices and systems to operate in their intended environment without causing or suffering frequency fratricide. The four components of EMSO are spectrum management, frequency assignment, policy implementation, and host-nation coordination. EMSO supports the six warfighting functions of command and control, intelligence, fires, movement and maneuver, protection, and sustainment through the full spectrum of war (U.S. Army, 2010).

Spectrum management strives to allocate and manage the use of spectrum to ensure that users' spectrum requirements are met without

Table A.1
EW Functional Divisions

Division	Details
Electronic attack	Electromagnetic jamming, electromagnetic deception, directed-energy and antiradiation weapons, expendables
Electronic protect	Spectrum management, EW hardening, emission control
Electronic warfare support	Threat warning, collection supporting EW, direction finding

creating frequency fratricide. Spectrum management consists of many tasks, such as evaluating and mitigating electromagnetic environmental effects, managing frequency records and databases, deconflicting frequencies, frequency interference resolution, allocating frequencies, and EW coordination (U.S. Army, 2010).

Frequency assignment involves the requesting and issuance of authorization to use frequencies for specific devices or systems, such as a combat net radio network, remotely piloted vehicles, or line-of-sight networks (U.S. Army, 2010).

The commercial and governmental bodies of the world use and depend on the EMS. Implementation of policies to regulate and coordinate its use among various users is vital to the effective and efficient use of this limited resource. At the global level, the International Telecommunications Union coordinates the use of spectrum every two to three years at the World Radio Communication Conference. In the United States, the Military Communications-Electronics Board is the main coordinating body for spectrum policies in DoD. The primary mechanism for allocating and enforcing proper use of the spectrum among DoD systems is the J/F 12 Spectrum Certification process. All systems and equipment that emit or receive Hertzian waves must submit a DD Form 1494, Application for Equipment Frequency Allocation (U.S. Army, 2010).

The EMS is partitioned and allocated to each nation, and each nation has sovereignty over its spectrum. Transmissions within a country are subject to that country's regulations and evaluation of potential interference with local users. The use of military and commercial systems inside host nations requires lengthy coordination and negotiations to reach approval and certification status (U.S. Army, 2010).

Information Assurance

IA is defined as

> Measures that protect and defend information and information systems by ensuring their availability, integrity, authentication,

confidentiality, and non-repudiation. This includes providing for restoration of information systems by incorporating protection, detection, and reaction capabilities. (U.S. Department of Defense Directive 8500.01E, 2007)

Distinction Between Information Assurance and Computer Network Defense

IA applies more broadly to information and information systems.[3] According to section 5.3 in U.S. Department of Defense Instruction 8500.2 (2003), CND constitutes the "operational component of IA." Some contend that the "fundamental difference between IA and CND is not in the context of friendly operations [but] in the context of enemy operations" (Stern, 2008). Nonetheless, it is often the case in the broad literature that IA and CND are used interchangeably. According to Stern (2008), this is also true for the Army.

IA is more general, in part because it includes many of the non-materiel aspects of securing the Army's portion of cyberspace (e.g., policy, compliance, best business practices). CND is more focused on the computer and network itself, as well as on operational capabilities to support missions to actively defend computers and networks, especially in real time or near real time. One viewpoint that reflects some operators is this: It is vital for the Army (e.g., TRADOC) to clearly distinguish the two.

> The Army must establish a distinction between Information Assurance and Computer Network Defense. . . . [T]his will also help facilitate the definition of roles and responsibilities between [network operations] and [information operations]. This will help to bring about an understanding for the need for both terms, how they integrate, and redefine them to line up with standard military doctrine. (Stern, 2008)

---

[3] This is relative to CND, which, according to JP 3-13, "involves actions taken through the use of computer networks to protect, monitor, analyze, detect, and respond to unauthorized activity within DOD information systems and computer networks" (U.S. Joint Chiefs of Staff, 2006a).

Public Affairs

According to JP 3-61, PA provides information to domestic and international audiences and contributes to global influence and deterrence of attacks (U.S. Joint Chiefs of Staff, 2010a). This guidance does not distinguish between adversaries and U.S. audiences; it includes both. Army Field Manual 46-1 says, "PA operations are directed toward U.S. forces and U.S. and international media" (HQDA, 1997b, p. 13). The target audiences may differ, but the consistency of messages is important to credibility. PA focuses on truth and credibility and must be fact-based.

Knowledge Management

Defining Knowledge Management

Knowledge management largely encompasses people, processes and technology; it uses information requirements and analysis to produce relevant information.

Aspects of knowledge management are discussed in doctrine for network operations, in particular (i.e., FM 6-02.71, *Network Operations*; HQDA, 2010b). The topic is covered more generally in operations doctrine (i.e., FM 3-0, *Operations*; HQDA, 2008a). In doctrine for network operations, it is called information dissemination management/content staging. This concept focuses on the technical side of knowledge management (i.e., systems, software, graphical user interfaces, shareportal construct, format, and access).

Encompassing these technical means is the "art of knowledge management." As described in FM 3-0 (2008), it is the effort to improve situational understanding (e.g., "paint the picture") sufficiently to aid decisionmaking. FM 3-0 sums it up best: "Effective knowledge management requires effective information management." We summarize knowledge management as follows:

> *Knowledge management is the art and science of improving situational awareness and preventing information overload to a sufficient*

degree to support actions and decisions dynamically across the entire organization.

Personnel Issues

There are personnel assigned knowledge management duties:

> The KM section reports directly to the chief of staff or executive officer. The section may contain the following positions: a KM officer, an assistant KM officer, a KM noncommissioned officer, and content management specialists. Section member duties and responsibilities depend on the number of Soldiers assigned to the section. This number also determines how many functions the section can accomplish. Not all positions described here may be authorized or required at a given echelon. (HQDA, 2008b, Chapter 2)

Obviously, the knowledge management function needs attention.[4] While the knowledge management officer does not work for the FA30, this person is relevant to IO tasks.

New Doctrine, Recent Decisions, and Ongoing Study Efforts

TRADOC Memorandum

In 2009, GEN Martin Dempsy, commanding general of TRADOC, authored a memorandum to the Army Vice Chief of Staff recommending a way forward for cyber operations, EW, and IO. In the memo, he concluded the following:

1. The cyber-EW-IO vocabulary in use today is adequate—but will become increasingly inadequate—to describe future challenges brought about by the rapid acceleration taking place in

---

[4] Elder (2010) notes, "Knowledge management is a key aspect of mission assurance. Knowledge Management is not given enough attention operationally or administratively; it will grow in importance to match the explosion in information which must be processed to develop useful knowledge."

commercial IT and adversaries' capabilities to use and adapt it to their purposes.

2. Future challenges require a new conceptual framework that spans three interconnected dimensions: contest of wills, strategic engagement, and the cyber/electromagnetic contest.

3. This framework is a way forward to enable the future force to meet the demands of full-spectrum operations in an operational environment characterized by complexity, rapid changes, and hybrid adversaries that will function in both major combat and IW scenarios.

Electronic Warfare Capability-Based Assessment

The EW CBA foreshadowed the TRADOC memo in that it explicitly recognized the limitations imposed by then–joint and Army EW policy that segregated EW as a separate field, precluding its combined consideration with SIGINT and CNO (TRAC, 2007). The result precluded taking full advantage of their synergies. Properly done, the interconnected framework should enable the use of such synergies.

Cyberspace Operations Concept Capability Plan

The cyber operations concept capability plan (TRADOC, 2010a) was the next step in exploring the new framework outlined in the Dempsey memorandum. The plan's intent is to

> develop a common understanding of how technological advancements transform the operational environment, how leaders must think about cyberspace operations, how they should integrate their overall operations, and which capabilities are needed. (TRADOC, 2010a)

The Army cyberspace concept capability plan (TRADOC, 2010a) views cyber operations as a means of leveraging cyberspace and the EMS and thereby prevailing in the cyber/electromagnetic contest. It defines four components for cyber operations:

- cyber situational awareness
- cyber network operations

- cyber warfare
- cyber support–enabling capabilities and approaches.

The Cyber/Electromagnetic CBA

Of the three dimensions described by Dempsey (contest of wills, strategic engagement, cyber/electromagnetic contest), an ongoing CBA is addressing the third. It intends to focus on "gaining and maintaining an advantage in the converging mediums of cyberspace and EMS"[5] (TRADOC, 2010a).

---

[5] Another CBA is being planned for the other dimensions—contest of wills and strategic engagement—that will focus on how commanders and staffs orchestrate and leverage information power to achieve their objectives.

Information Operations in Doctrine

Currently, IO is defined as

> the integrated employment, during military operations, of information-related capabilities in concert with other lines of operation to influence, disrupt, corrupt, or usurp the decision-making of adversaries and potential adversaries while protecting our own. (Gates, 2011)

IO grew out of emergent thinking in the 1990s that was concerned with what, exactly, information-age warfare would look like. What began in doctrine as C2W (CJCSI 3201.01, 1996) was part of a broader construct—information warfare—and ultimately evolved into information operations. C2W was built around an emphasis on the role of information and information technology in friendly and adversary OODA loops (Coran, 2004) and was based on the reasonably sound notion that focused efforts to corrupt or slow an adversary's OODA loop (while protecting that of the United States) would lead to operational advantages.

In 1996, C2W was defined in joint doctrine as

> the integrated use of psychological operations (PSYOP), military deception, operations security (OPSEC), electronic warfare (EW), and physical destruction, mutually supported by intelligence, to deny information to, influence, degrade, or destroy adversary C2 capabilities while protecting friendly C2 capabilities against such actions. (CJCSI 3201.01, 1996)

The parallels between the definitions of C2W in 1996 and IO in current doctrine are striking. While the concept and practice of IO has eclipsed its origins in C2W, the definition has not yet done so.

Interestingly, also in 1996, the Army released doctrine for IO, FM 100-6. That publication explicitly and intentionally considered IO to be broader than just C2W:

> Joint Pub 3-13.1 states that beyond the five fundamental elements of C2W "other capabilities in practice may be employed as part of C2W to attack and protect." The Army recognizes that C2W is the joint reference point for IO when working with the joint staff and other services in the realm of IW. However, the Army interprets this new paradigm more broadly and recognizes the more comprehensive integration of other information activities as fundamental to all IO; hence the term operations, which includes specifically C2W, CA, and PA. (HQDA, 1996, p. 3-0, fn 1)

Army doctrine in 1996 offered this definition of IO:

> continuous military operations within the military information environment that enable, enhance, and protect the friendly force's ability to collect, process, and act on information to achieve an advantage across the full range of military operations; information operations include interacting with the global information environment and exploiting or denying an adversary's information and decision capabilities. (HQDA, 1996, Glossary-7)

While still sharing the focus of C2W on adversaries and OODA loops, this 1996 Army definition was much broader (and discussed more broadly in the text) than the first joint definition of IO, released in 1998.

In 1998, the first edition of JP 3-13, *Information Operations*, was released. The scope was once again considerably narrowed when this first edition of JP 3-13 defined IO as "actions taken to affect adversary information and information systems while defending one's own information and information systems" (U.S. Joint Chiefs of Staff, 1998).

This brought the concept squarely back in line with its roots in C2W, roots that continue to leave a significant legacy in the doctrine, if not in the practice, of IO.

The 2003 Information Operations Roadmap

The year 2003 was big for IO. October of 2003 saw the release of Secretary Rumsfeld's Information Operations Roadmap (classified at that time), and November of 2003 saw the release of FM 3-13, *Information Operations*, replacing FM 100-6. The 2003 IO Roadmap was a response to the 2001 Quadrennial Defense Review identifying IO as one of six critical operational focal points for transformation in DoD. The goal of the roadmap was to advance "information operations as a core military competency" (DoD, 2003, p. 1). The roadmap advocates a "common understanding of IO" and offers a definition to be included in joint doctrine and DoD directives on IO:

> The integrated employment of the core capabilities of electronic warfare, computer network operations, psychological operations, military deception, and operations security, in concert with specified supporting and related capabilities, to influence, disrupt, corrupt or usurp adversarial human and automated decision making while protecting our own. (DoD, 2003, p. 11)

This is the same definition that appears in the 2006 version of JP 3-13 and the current joint definition.

Even in 2003, there was concern about the lack of a shared understanding of IO in DoD. The roadmap notes, "The Services, combatant commands and Agencies do not have a common understanding of IO" (DoD, 2003). One of the goals of the roadmap was to move toward a common understanding of IO. That remains an important goal.

Field Manual 3-13 (2003)

In the November 2003 version of FM 3-13, *Information Operations*, IO was defined as

> the employment of the core capabilities of electronic warfare, computer network operations, psychological operations, military deception, and operations security, in concert with specified supporting and related capabilities, to affect or defend information and information systems, and to influence decisionmaking." (HQDA, 2003)

This definition was new to Army doctrine at that time but was "current with joint initiatives" (i.e., the IO Roadmap). This was a break from previous Army doctrine and saw the breadth implied in the 1996 version of FM 100-6 locked down to correspond with the direction laid out in the IO Roadmap. The only difference between this definition and the roadmap definition (still current today) is at the very end; what was in 2003 "to affect or defend information and information systems, and to influence decisionmaking" is now "to influence, disrupt, corrupt or usurp adversarial human and automated decision making while protecting our own."

Joint Publication 3-13 (2006)

The 2006 version of JP 3-13 superseded the 1998 version and made changes such that it "[a]ligns joint information operations (IO) doctrine with the transformational planning guidance as specified by the 30 October 2003 Department of Defense Information Operations Roadmap" (U.S. Joint Chiefs of Staff, 2006b, p. iii). The five core related capabilities remained unchanged, though some shuffling, additions, and changes were made in the supporting and related capabilities. "Strategic communication" appears in joint doctrine for the first time, with a list of capabilities that can support strategic communication (PA, defense support to public diplomacy, and IO) and a short discussion of the importance of an integrated information strategy (U.S. Joint Chiefs of Staff, 2006b, p. I-10).

Field Manual 3-0 (2008)

In February 2008, the Army released a revision of FM 3-0, *Operations* (HQDA, 2008a). This document made some fairly significant doctrinal changes to how the Army views and treats information. First, information was elevated to be one of the elements of combat power, alongside leadership, protection, movement and maneuver, intelligence, fires, sustainment, and command and control (HQDA, 2008a, p. 4-1). Though an element of combat power, information was not considered a "warfighting function," a decision that is once again under consideration at this time.

In Chapter 7, "Information Superiority," the 2008 version of FM 3-0 decomposed information into five distinct Army information tasks: information engagement, C2W, information protection, OPSEC and MILDEC (HQDA, 2008a, p. 7-2).

Beyond simply specifying these separate tasks, the biggest practical change was in how responsibility for these efforts was assigned to staffs. Gone was the assignment of all these traditional IO roles to a single IO staff component.

Consisting of a blend of PA, FA30, and MISO personnel, information engagement is the staff responsibility of the G-7. EW and CNO form the C2W cell under the fires support coordinator. Information protection, formerly IA, remains with the G-6; OPSEC belongs to G-3 Protect, and MILDEC to G-3 Plans (Rosin, 2009). This, too, remains hotly contested.

The 2009 Attempt to Revise FM 3-13

In February 2009, the IO Proponent Office at the TRADOC Combined Arms Center released a draft of FM 3-13, provisionally titled *Information*, for Army-wide staffing (Henderson, 2009, p. 4). The draft was synchronized with the 2008 version of FM 3-0, and listed the same five information tasks: information engagement, C2W, information protection, OPSEC, and MILDEC. Where it broke new ground was in specifying three operational challenges that commanders will face in full-spectrum operations.

These challenges are (1) maintaining the trust and confidence of home and allied publics while gaining the confidence and support of local publics and actors; (2) winning the psychological contest of wills with adversaries or potential adversaries; and (3) winning the contest for the use of information technology and the EMS (Whisenhunt, 2009, p. 29).

The draft doctrine was intended to expand "the Army's approach to the use of Information beyond 'information operations'" (Henderson, 2009, p. 4). It echoed the contentious staffing changes for IO and related capabilities that were advocated in the 2008 draft.

The 2009 draft was lauded by some and reviled by some. As the director of the IO Proponent Office noted, "Responses to the Initial Draft have ranged from publish now to start over" (U.S. Army Information Proponent Office, 2009a). "Start over" won out, with the draft first delayed by TRADOC's GEN Martin Dempsey and then canceled altogether (Gould, 2009).

Proposed Changes to FM 3-0 in 2010

Although FM 3-0 was revised in 2008, there is a new draft revision circulating as of this writing. The draft's revision notes indicate significant changes in two areas: the mission command warfighting function and information tasks. The April 30, 2010, circulating draft revision of FM 3-0 retains the 2008 concept of "information tasks" but changes from five Army information tasks to only two: inform and influence activities, and cyber-electromagnetic activities. These tasks are explicitly integrated into full-spectrum operations:

> In an environment saturated by information, messages, themes, and actions are inextricably linked. Effective Full Spectrum Operations require integrated themes and messages, synchronized with actions. The most powerful message that Soldiers send is their actions on the ground. (HQDA, 2010a, p. 7-1)

One of the significant changes in the new draft is the creation of the "mission command" warfighting function, and one of the significant discussions surrounding the drafting process is whether "information," in addition to being an element of combat power, should be a warfighting function. The drafters conclude that no, information should not be a warfighting function; instead, they assign many of the functions that would go into an information warfighting function to the mission command warfighting function. This affects the proposed staffing arrangement, in which an inform and influence section reports directly to the Army Chief of Staff, as does the cyber-electromagnetic section (HQDA, 2010a, Sec 7-53 and 7-54). Figure B.1 contrasts the elements of combat power in the 2008 version of FM 3-0 with the proposed revision.

2010 PSYOP Name Change

In June 2010, the name of the PSYOP function was changed to military information support operations, or MISO (Csrnko, 2010). Presumably, this was done because of problems stemming from the imputation of evil manipulation that adhered to PSYOP. Admiral Eric T. Olson, commander of U.S. Special Operations Command, indicated

Figure B.1
The Elements of Combat Power in FM 3-0, 2008 and 2010 Proposed Rewrite

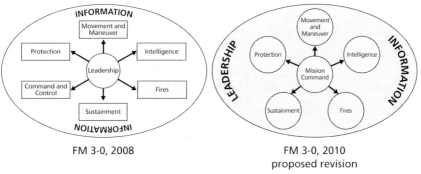

FM 3-0, 2008

FM 3-0, 2010
proposed revision

RAND *MG1113-B.1*

that this was more than just a name change; it "will be a complete change in organization, practice, and doctrine" (Paddock, 2010, p. 1).

While this discussion is not directly related to IO, it is certainly relevant. MISO is one of the core capabilities of traditional IO, and it is the central military capability for influence messaging. Significant changes in the organization, practice, and doctrine for MISO could have a significant impact on this area's relationship to IO.

Expected Action in Late 2010 and 2011

IO appears to be a moving target. There was a subpanel for IO for the 2010 Quadrennial Defense Review, which reportedly reexamined the definition of IO (Kuehl, 2009, p. 6). As of this writing, there are several IO-related studies under way and expected to be released in late 2010 or 2011. These studies include the Strategic Communication CBA, the Joint IO Force Optimization Study, at least two MISO CBAs (one for the reserve component and one for the active-duty component), and the strategic communication and IO front-end assessment ("DoD Launches POM-12 Study on Info Ops, Strategic Communication," 2010).

In January, 2011, Secretary of Defense Robert Gates issues a memorandum outlining a revised definition of IO, with a greater focus on integration. He stated that the definition in effect when this research was conducted placed "too much emphasis on core capabilities" and supported the "notion that the core capabilities must be overseen by one entity. Joint doctrine now defines IO as

> the integrated employment, during military operations, of information-related capabilities in concert with other lines of operation to influence, disrupt, corrupt, or usurp the decision-making of adversaries while protecting our own. (Gates, 2011)

International IO Developments

Given the confusion in the United States with regard to IO, perhaps it is worth considering how such operations are defined by various allies. The NATO definition is not that much different from U.S. definitions, including the residual C2W focus on information processes and protecting those of the United States, but it does explicitly emphasize seeking effects on the will and understanding and does specifically include targets beyond an adversary:

> Coordinated actions to create desired effects on the will, understanding, and capability of adversaries, potential adversaries and other approved parties in support of Alliance overall objectives by affecting their information, information-based processes and systems while exploiting and protecting one's own. (NATO, 2007)

Similarly, the UK definition is very much in the C2W tradition, targeting only adversaries or potential adversaries:

> Coordinated actions undertaken to influence an adversary or potential adversary in support of political and military objectives by undermining his will, cohesion and decision-making ability, through affecting his information, information-based processes and systems while protecting one's own decision-makers and decision-making processes. (UK Ministry of Defence, 2002)

The Australian Army, however, seems to be moving away from traditional IO constructs, replacing *information operations* with *information actions* and considering this definition for the latter:

> Actions conducted to influence target audiences in order to achieve understanding, acceptance, and support of our actions and aims, and to diminish the quality and speed of the adversary's decision making, while maintaining our own, to achieve decision superiority. (Nicholas, 2008)

The Australian Army does not view retention of the term *information operations* as critical, but it does assert that retention of the underlying influence concept is paramount (Nicholas, 2008).

Congressional Pressure Concerning IO and Strategic Communication

Finally, part of the impetus for the recent flurry of activity in the realm of IO and strategic communication is congressional pressure. There has recently been concern that DoD IO and strategic communication have been poaching on the Department of State's traditional public diplomacy mission and that DoD cannot adequately account for the expenditures and effects of these operations (Armstrong, 2009). Congress cut hundreds of millions of dollars from the 2010 defense budget for IO (Gertz, 2009). Some defense officials continue to worry that apparent confusion regarding IO will give congressional appropriators further excuses to cut funding for IO (Ambinder, 2010, p. 3).

Issues Regarding Information Operations as Integration, Advocacy, and/or a Capability

In this appendix, we identify a series of issues that relate to the question of whether IO should be about integration, advocacy, and/or a capability in its own right. Key considerations include the following:

- traditional compared with nontraditional military activities
- access to the commander
- full integration and coordination rather than deconfliction
- combined arms integration.

Core Kinetic Activities Compared with Information Warfare

One of the challenges that IO has faced (and no doubt part of the impetus for the initial construct) is that the elements of information warfare are not embedded in the development processes that all soldiers go through. Every Army officer understands the concept of fires, the relevance and conduct of maneuver, the principles of combined arms, the implications of different kinds of terrain, and the importance of logistics. What every officer does not necessarily intuitively understand is the relationship of cyberspace to the battlespace, how spectrum use is managed, and the cognitive effect on the target audience that will likely result from a certain scheme, maneuver, or set of rules of engagement.[1] As long as information capabilities and effects remain outside the core of military training, there will be challenges in integrating them with core kinetic capabilities.

In addition to the absence of IO from the Army's military development processes, planning and executing IO differs substantially from kinetic military activities. Consider first that IO involves a much larger space of operations, in terms of both the variety of possible effects and the capabilities available to produce them (Allen, 2007). The range of both the time and space available to IO can be very broad most of the time (Allen, 2007).[2] The global information environment is, of course, global, but it contains numerous local environments of grave importance to operations. Bits can fly in microseconds, but long-term influence objectives can take years to realize. Building on this theme, most IO effects do not result from direct action, but create indirect effects, sometimes second- or third-order effects (Allen, 2007). Finally, because of varying timelines and the lack of direct effects, getting feedback on IO effects can be particularly challenging (Allen, 2007). Traditional battle damage assessment will not work for most information effects, and measures of effectiveness can be difficult to conceive—and even more difficult to connect to information activities.[3] These facts argue strongly for the need for an information advocate or proponent as part of the battle staff.[4]

---

[1]　Elder (2010) adds that "most officers understand the use of force to defeat an adversary physically; however, few are comfortable with the concepts underlying the use of military forces to alter adversary behaviors without engaging in combat."

[2]　IO timelines can be quite short if a situation develops that requires the quick deployment of either a PA message or MISO/civil affairs resources. The same is true if there is collateral damage. An associated difficulty is that messages and themes (generally) must be coordinated with preexisting guidance from higher echelons, and permission must be given for new messages and themes (to ensure the deconfliction of messages).

[3]　For example, Helmus, Paul, and Glenn (2007, p. 47) state that "many Iraqi soldiers surrendered at the outset of OIF. Was this due to the PSYOP leaflets dropped instructing them to do so? Was it instead due to the impact of the coalition's massive military might? Were there other causes?"

[4]　This is the function of the S7, if not the effects coordinator, fire effects coordination, etc.

To some extent, the lack of an IO perspective in Army development processes is a generational challenge (see Helmus, Paul, and Glenn, 2007). Today's young officers grew up embedded in cyberspace and have a much better intuitive feel for the domain and its role in warfighting. Those who have served at the company and battalion levels in Iraq and Afghanistan have a profound sense of the importance of informing, influencing, and persuading combat and noncombat audiences; understand that such efforts are central to the outcomes of certain kinds of operations; and realize the importance of including such considerations in operational design. As this generation matures and rises through the ranks, and as doctrine, training, and professional military education evolve to put additional emphasis on information tasks and capabilities, more and more senior commanders will have an information mindset. Eventually, IO will be a part of the core military tradition. Until that time, if the Army values IO, it may need to include some kind of information advocate at the right hand of the commander, ensure that desired information goals are identified in planning guidance, and ensure that the information effects of all operations (including the kinetic) are recognized and coordinated.[5]

Access to the Commander

Another issue that is relevant to the question of IO as integrator, advocate, or capability concerns access to the commander. Because the commander is unlikely to be completely expert in the various capabilities available to affect the information, cognitive, or cyber domains, he or she needs access to such expertise. What is the best way to provide this access?

---

[5] As suggested by Elder (2010), "One can argue that, rather than an information advocate, a commander requires a strategic communications 'devil's advocate' to offer the commander an assessment of how actions will be perceived, regardless of the accompanying messages." One aspect of IO is an assessment of how operations will shape the information environment, even if no "planned" influence or inform operations are conducted or directly intended.

The current IO construct places the IO officer (as the integrator) above the IO capability personnel and between them and the commander and the rest of the staff. As mentioned, IO officers are currently not required to have special expertise in any of the IO capabilities, let alone all of them. While a smart and capable IO officer might recognize his or her own shortcomings and the needs of the commander, and while that officer might regularly bring, for example, a PSYOP (now MISO) officer to higher staff meetings or to meet with the commander, another IO officer who simply follows the chain of command and protects his or her own "ricebowl" might not. When this happens, the commander may have a difficult time accessing relevant expertise (Helmus, Paul, and Glenn, 2007). As a recent *Military Review* article reports,

> PSYOP officers rarely talk directly to commanders. Communication usually goes through the IO officer or strategic communication officer to the operations officer or chief of staff and then to the commander. Too often, the commander talks directly to foreign populations without the aid of PSYOP units. (Rohm, 2008)

PA manages to avoid the extra bureaucratic layer (the IO layer) between its capability and the commander in a way that the core IO capabilities cannot. As the commander's spokesperson, the PAO has a special relationship with the commander, serving as a special staff officer with direct access to the commander (as does, for example, the commander's staff judge advocate).[6] This seems like the perfect organizational location and relationship for an information advocate. The Air Force and the Army experimented with placing PAO personnel in the influence cell with good results (Elder, 2010). The 1/25 SBCT, for example, used a PAO for quite some time. That person worked in a support role in the joint information bureau at the Army's National Training Center and the Joint Readiness Training Center and coordinated stories with the corps PAO on a regular basis in garrison. (See Appendix E for a more detailed discussion.) In most units (brigade and

---

[6] The special staff officer is a traditional way to provide the commander with direct access to special expertise.

below, but sometimes division), there is no PAO in garrison; the PAO is attached on deployment only.

When deciding whether the role of IO should be integration, advocacy, or both, it is necessary to keep in mind the extent to which commanders have access to the expertise they need. Depending on the role chosen for IO in the future, based on the three core questions here, future IO officers may need different training, or they may need to come from specific capability backgrounds (e.g., MISO, EW) to adequately serve the new vision.

Full Integration and Coordination Rather Than Deconfliction

In current doctrine, IO integrates the five core capabilities (along with the related and supporting capabilities). To what extent do these capabilities need to be fully *integrated*, and what is the minimal level of deconfliction required? Whether integration or deconfliction, how does that differ from the combined arms deconfliction of other military capabilities?

There are strong arguments for the integration of certain IO capabilities. For example, the boundary between CNO over wireless networks and EW is not always clear, and operators in those two specialties had better be fully integrated to avoid duplication of effort or working at cross-purposes. Similarly, both MISO and PA produce messages with intended effects. If these areas are not fully integrated, messages will be contradictory and possible synergies will be lost.

What about across the content-systems divide? From some perspectives, EW and MISO require only deconfliction. EW can support MISO by jamming competing broadcasting sources, for example. Here is how one IO officer characterized the integrative relationship between EW and MISO (formerly PSYOP; quoted in Paul, 2008):

PSYOP: "Don't jam that."
EW: "Got it."

MISO and CNO do have some integrative synergies. Imagine, for example, that a MISO objective involves influencing a specific adversary decisionmaker and that such influence would be facilitated by placing electronic messages on that individual's computer. Perhaps MISO and CNO working in concert could accomplish that. Similarly, CNO technical experts may succeed in penetrating an adversary network. If their objectives for that penetration involve inserting other than technical content, then they should certainly integrate with MISO personnel first to ensure that the content they inject is optimally designed and does not conflict with other content-based initiatives.[7]

Combined Arms Integration

Perhaps the broader issue is whether the extent to which the IO capabilities require integration with each other differs from the integration needs of traditional combined arms capabilities. As BG (ret.) Huba Wass De Czege (2008) notes,

> The notion of supporting and related IO capabilities has always struck me as strange as labeling artillery, intelligence and other supporting or related branches as supporting or related Infantry capabilities. In the Army's concept of combined arms operations, the various branches, capabilities and competencies of the Army are all related and mutually supporting when they serve a common purpose beyond the technical purpose for which they are differentiated. The common practice is to combine capabilities necessary and sufficient to achieve the objective of the taskforce.

In fact, in some cases, there is more need to integrate IO capabilities with other combined capabilities than with other IO capabilities. For example, MISO influence efforts may need to be tightly integrated with a scheme of maneuver or with the actions and behavior of troops. Otherwise, they risk becoming false and losing credibility. Failure to

---

[7]   Rarely are these operations conducted independently.

integrate MISO with capabilities like EW or CNO could, at worst, result in a message not being disseminated.

Common Electronic Warfare and Electromagnetic Spectrum Operations Tasks and Overlaps

Table D.1
Electronic Warfare Tasks

EW No.	Electronic Warfare Task
EP 1	Protect friendly personnel, equipment, systems, information, and facilities from adverse EW effects (electromagnetic hardening).
EP 2	Protect the use of the EMS, including spectrum management and RF deconfliction (for radar, communication, warheads, relays, sensors, etc.).
EP 3	Conduct electromagnetic hardening.
EP 4	Coordinate and modify emission control measures.
EP 5	Coordinate EW reprogramming.
EP 6	Conduct friendly EW strike warning.
ES 1	Search for, intercept, identify, locate, classify, and display sources of intentional and unintentional radiated electromagnetic energy.
ES 2	Integrate, analyze, and fuse collected data and information to provide targetable intelligence in support of EW.
ES 3	Exploit intentional and/or unintentional radiated electromagnetic emissions in support of immediate EW operations.
EWI 1	Describe and depict the EME.
EWI 2	Maintain electronic order of battle.
EWI 3	Recommend dynamic adjustment of EW resources.
EWI 4	Evaluate, integrate, analyze, and interpret operational information as it relates to EW operations.

Table D.1—Continued

EW No.	Electronic Warfare Task
EWI 5	Establish, populate, update, and propagate EW information into the COP.
EWI 6	Prepare and develop the EW annex to include support to fires.
EWI 7	Plan scalable EW effects (lethal and nonlethal fires).
EWI 8	Establish EW target priorities.
EWI 9	Coordinate EW support to targeting.
EWI 10	Coordinate and synchronize EW operations between/across all phases with joint, interagency, intergovernmental, and multinational organizations.
EWI 11	Collaboratively assess achievement of planned EW effects.
EWI 12	Collaboratively identify and assess the implications of unintended EW effects.
EWI 13	Evaluate EW measures of effectiveness, assess measures of performance, and conduct battle damage assessment to determine effects achieved.
EA 1	Jam adversary electromagnetic capabilities.
EA 2	Conduct EA with directed energy (including pulse and high-power microwave).
EA 3	Conduct antiradiation operations to destroy, degrade, or neutralize enemy systems.
EA 4	Conduct electronic deception operations.
EA 5	Conduct defensive EA to protect personnel, equipment, systems, information, and facilities (self-protection).
EA 6	Employ EW to destroy, neutralize, or suppress EW or chemical, biological, radiological, or nuclear capabilities.
EA 7	Use obscuration to defeat enemy use of the EMS (visual, infrared, millimeter wave).

NOTE: EP = electronic protect. EWI = electronic warfare integration. EA = electronic attack.

Table D.2
Electromagnetic Spectrum Operation Tasks

EMSO No.	EMSO Task
SM 1	Plan the use of the EMS for all spectrum-dependent devices (EMSO mission planning) (battalion, brigade, division, corps, task force, theater, and installation levels).
SM 2	Conduct electromagnetic interference analysis (soldier to installation).
SM 3	Provide EME information in either a networked or stand-alone mode (build EME COP).
SM 4	Perform modeling and simulation of the EME via user-selected data fields of the impact of the EME on projected EMS use plans.
SM 5	Monitor and use spectrum COP information in support of full-spectrum operations (company to installation level).
SM 6	Prioritize spectrum use based on commanders' guidance.
SM 7	Utilize EW reprogramming during the nomination, assignment, and deconfliction processes.
SM 8	Generate and distribute signal operating instructions and joint communications-electronics operations instructions (brigade, division, corps, theater, and installation levels).
SM 9	Create, import, export, edit, delete, display, and distribute the Joint Restricted Frequency List.
SM 10	Access and use EMSO technical data.
SM 11	Manage, store, and archive EMS use data (frequency management work history).
FA 12	Assign frequencies within the operational parameters of the emitter and available resources.
FA 13	Obtain requests and provide EMS resources to requesting unit (battalion, brigade, division, corps, task force, theater, and installation levels).
FA 14	Import satellite access authorization.
HN 15	Utilize host-nation comments in the spectrum nomination and assignment process.

NOTE: SM = spectrum management. FA = functional area. HN = host nation.

Table D.3
Overlapping EW and EMSO Tasks

EW Number \ EMSO Number	SM 1	SM 2	SM 3	SM 4	SM 5	SM 6	SM 7	SM 8	SM 9	SM 10	SM 11	FA 12	FA 13	FA 14	HN 15
EP 1	■	■	■	■	■					■		■	■		■
EP 2	■	■	■	■	■										■
EP 3		■		■						■					■
EP 4		■	■	■	■							■			
EP 5	■														
EP 6			■											■	
ES 1					■										
ES 2					■										■
ES 3															
EWI 1						■								■	■
EWI 3		■										■			
EWI 4	■		■			■	■						■		■
EWI 5	■														
EWI 6													■		
EWI 7	■			■		■				■			■		■
EWI 9	■														
EWI 10															
EWI 11	■						■					■	■		■
EWI 13															
EWI 14															
EWI 15		■													
EWI 16															

NOTE: EP = electronic protect. ES = electronic support. EWI = electronic warfare integration.

Discussion: Information Operations in the 1/25 Stryker Brigade Combat Team

This appendix examines the organization of a particular SBCT for the purposes of conducting IO. IO, as conducted by the 1/25 SBCT by the early 2000s, experimented with center-of-gravity analysis and different ways to influence decisionmakers, either by degrading decisionmaking or by confirming decisions made by the "decisionmaker." Each member of the "effects" community in the 1/25 was trained (in some fashion and to a greater or lesser degree) on lethal and nonlethal effects. Training was also provided to battalion commanders and staffs.

As discussed in Chapter Eight, in our view, establishing the G-7 as the information engagement staff and assigning to it influence messaging capabilities will only serve to marginalize the influence component of IO. After the commander and the chief of staff, the J/G-3 has much control. If one wants something (like influence) to be treated as though it is important and has priority, putting it outside or below J/G-3 is *not* the way to accomplish that. Further, separating information engagement from operations excuses operations from the need to integrate with engagement and engagement from the need to integrate with operations, making the marginalization complete. Only a commander firmly committed to maximizing influence information effects would be able to effectively integrate influence with maneuver and fires under this staffing scheme, and then only through constant vigilance.

In the 3/2 SBCT, the IO officer was the S7 (training), and the S5 (plans) handled civil-military operations. In the 1/25, both were aligned under the effects coordinator. Thus, in the 3/2, the S7 was somewhat marginalized. As each component was separated with different "bosses," the integration function was lost. Housing all the core IO personnel together, with the weight of fires, improved integration and elevated the importance of IO functions on the 1/25 staff.

IO in 1/25 SBCT was organized along the coordinated effects concept. Although effects-based operations are out of favor in Army terminology, the organizing principle is useful for this discussion. The 1/25, in conjunction with the Land Information Warfare Activity and Fort Sill, and the given Modified Table of Organization and Equipment, was organized as follows (in descending order by rank/command): brigade commander, effects coordinator (also the field artillery battalion command, principal adviser to the brigade commander on effects), deputy effects coordinator (formerly the brigade fire support officer, in charge of the Fires and Effects Cell on the brigade staff), IO coordinator (in charge of the Non-Lethal Effects Cell). The Fires and Effects Cell and the Non-Lethal Effects Cell were merged into the Effects Coordination Cell, which oversaw integration, coordination, planning, offense, and defense. It consisted of an IO coordinator (an FA30 IO officer, a major), a senior PSYOP noncommissioned officer (E-8, principal planner and coordinator of attached PSYOP assets), a civil affairs officer (FA39, principal planner and coordinator of attached civil affairs assets), an EA officer (35G SIGINT/EW planner and coordinator for EA), a tactical intelligence officer (35D planner and integrator of IO and intelligence), a brigade operational law team (including officers and noncommissioned officers, who advised the commander but also ensured that PA/PSYOP/civil affairs efforts were legal).

Because there was no PAO assigned to the brigade, this was an additional duty assigned to the cell (PAOs or PA deputies were attached only upon deployment). Because of the limitations of equipment and staff, the IO section focused on the following IO functions: OPSEC, PSYOP, MILDEC, EA, physical destruction, PA, and civil affairs in coordination with the fires effects coordination cell. Because there were no "integrators" at the battalion and company levels, fire support officers were cross-trained in both lethal and nonlethal effects. In planning and orders writing, the Effects Coordination Cell produced an annex; IO was an appendix and each IO function was a tab. An important note on how this worked: each "effects" person, at all echelons, had a primary job (for example, the EA officer coordinated with other staff members on all electronic issues), but all positions required broad-based thinking about IO and the information environment. It was important that each effects staff member or operator had IO training so that they could integrate IO at all levels and retain perspective on IO despite their narrower job functions.

In summary, IO has different meanings and is conceived of and practiced differently at different levels. At strategic and joint levels, IO is more conceptual. At brigade-and-below levels, IO is more practical. The separation between IO as a concept and IO in practice is one of capacity, execution, and resources available.

Strategic IO concepts are embedded into plans as operations orders are pushed to and refined at lower and lower organizational levels. But at the tactical level it is difficult for small units and soldiers to translate IO concepts into practice, even though each and every soldier is in effect a conveyer of an IO theme or message. Despite its importance, IO as a concept does not neatly translate into tactical-level planning, processes, systems, or operations, although its value at the tactical level is immeasurable. For a more detailed discussion, see Paschall (2005).

Proposals for Navy Cyber Career Paths and Pipelines

Cyber career paths have been outlined for Navy officers, enlisted personnel, and civilians. This appendix outlines the Navy's proposals. The purpose is to provide an example of a cyber career path.

Information Dominance Corp

The Navy has combined its OPNAV N6 and N2 staffs to form a corps of information dominance professionals to better manage these personnel and achieve unity of effort. This change began in 2009. The OPNAV N2/N6 (Deputy Chief of Naval Operations for Information Dominance) and his staff lead this corps.

The stated goal of the corps is to create empowered professionals who can lead in the effort to transform the Navy and provide it with a position of prominence. The personnel sought include those who are skilled, inspired, and expert.

This corps of officers, enlisted personnel, and civilians is intended to include a cadre of cyber personnel. According to the Chief of Naval Personnel Public Affairs Office (2010),

> The Information Dominance Corps will create a cadre of information specialists, who come with individual community identities and unite to be managed as a corps, developed as a corps, and to fight as a corps . . . The Information Dominance Corps will consist of more than 44,000 active and Reserve Navy officers, enlisted and civilian professionals who possess extensive

skills in information-intensive fields to develop and deliver dominant information capabilities in support of U.S. Navy, Joint and national warfighting requirements. These fields include information professional officers, information warfare officers, naval intelligence officers, meteorological and oceanography officers, space cadre officers, aerographer's mates, cryptologic technicians, intelligence specialist, information systems technicians and civilian personnel.

As enumerated in Dorsett (2010, slide 14), the professionals in the Information Dominance Corps stem from existing and new personnel groupings, including

- meteorology and oceanography
- space
- intelligence
- information technology
- information warfare.

The Navy claims that "[t]he Information Dominance Directorate and the Fleet Cyber Command will allow the Navy to better staff, train and equip forces for cyber and information operations." There is speculation that these efforts will provide "a bigger role for Navy cryptologists in cyber operations" (Brewin, 2009).[1]

Career Paths for Commissioned Officers

Figure F.1 shows the notional career path for a Navy cyber officer as presented by U.S. Navy Personnel Command.

---

[1]   See quotes from ADM Gary Roughhead in Brewin (2009).

Figure F.1
Notional Career Path for Navy Cyber Officer

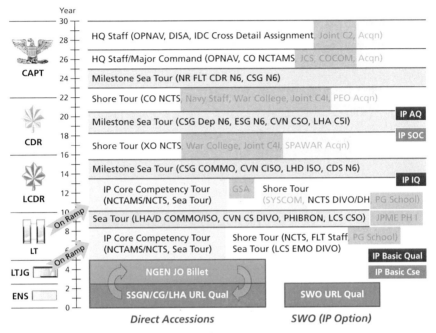

SOURCE: Barrett, Schroeder, and Carmickle, 2010.
RAND MG1113-F.1

Limited-Duty Officers

The Navy plans to utilize limited-duty officers (LDOs) for cyber operations. Traditionally, a limited-duty officer is chosen based on a specific skill set, e.g., medical professionals. Their assignments are focused exclusively in their skill set. In addition, LDOs may be limited in advancement, e.g., to captain. See Figure F.2.

Figure F.2
Navy Cyber Limited-Duty Officer Notional Career Path

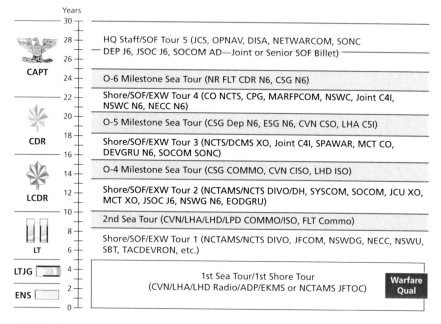

SOURCE: Barrett, Schroeder, and Carmickle, 2010.
RAND MG1113-F.2

Navy Cyber Warfare Officer/Engineer Career Path

The Navy is planning a position for technical experts in cyber operations similar to that of nuclear engineers (see Figure F.3). Accordingly, the cyber warfare officer (code 1810) will require a computer science, electrical engineering, systems engineering, or computer engineering degree for entry.

> The Cyber Warfare officer will serve in a probationary contract. If officers perform well, they can stay in the Navy via transfer or a civilian hiring process. (Kelsall, 2009)

Figure F.3
Navy Cyberwarfare Engineer Career Pipeline

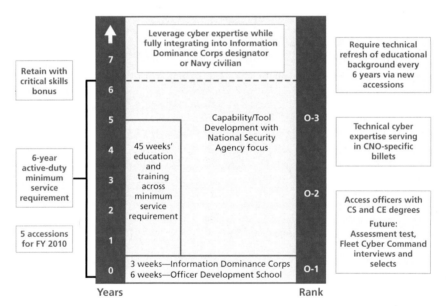

SOURCE: Barrett, Schroeder, and Carmickle, 2010.
RAND *MG1113-F.3*

Career Path for Navy Cyberwarfare Warrant Officer

There is currently a draft plan for a warrant officer (code 743X). It intends to be a vehicle for civilian to military transitions. Specific jobs include cyber security IA operator, cyber IA systems manager, and IA systems engineer (see Figure F.4).

Figure F.4
Navy Cyberwarfare Warrant Officer Career Pipeline

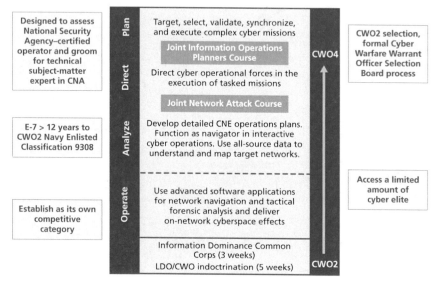

SOURCE: Barrett, Schroeder, and Carmickle, 2010.
RAND *MG1113-F.4*

Bibliography

AFDD—*see* Air Force Doctrine Document.

Air Force Doctrine Document 5, Information Warfare, second draft, October 1996, p. 27.

Allen, Patrick D., *Information Operations Planning*, Boston, Mass.: Artech House, 2007.

Ambinder, Marc, "Pentagon: Information Ops 'Plagued with Confusion,'" *The Atlantic*, June 15, 2010. As of December 6, 2010:
http://www.theatlantic.com/politics/archive/2010/06/pentagon-information-ops-plagued-with-confusion/58136/

Armistead, Leigh, ed., *Information Operations: Warfare and the Hard Reality of Soft Power*, Washington, D.C.: Brassey's, 2004.

Armstrong, Matt, "House Appropriations Concerned Pentagon's Role in Strategic Communication and Public Diplomacy (updated)," July 29, 2009. As of December 6, 2010:
http://mountainrunner.us/2009/07/public_diplomacys_combat_boots.html

Axelband, Elliot, RAND Corporation, "CERDEC Trip Report—Boundaries Between I/0, EW, Cyber, SIGINT, etc.," communication with Isaac R. Porche III, July 16, 2010.

Barrett, CAPT Danelle, CDR Julie Schroeder, and LCDR Bobby Carmickle, U.S. Navy Personnel Command, "Information Professional Community Brief," March 2010.

Beebe, Ken, "Misinformed on IO," letter to the editor, *Armed Forces Journal*, November 2009.

Boessenkool, Antoine, "Uncoordinated EW Procurement Creates Combat Issues," *DefenseNews*, November 11, 2009. As of December 6, 2010:
http://www.defensenews.com/story.php?i=4372242

Boyd, Curtis D., "Army IO is PSYOP: Influencing More with Less," *Military Review*, May–June 2007, pp. 67–75.

Brewin, Bob, "Navy to Form Directorate to Manage Info, Cyber and Space Capabilities," Nextgov, September 3, 2009. As of January 17, 2011: http://www.nextgov.com/nextgov/ng_20090903_3144.php

Brown, George C. L., "Do We Need FA30? Creating an Information Warfare Branch," *Military Review*, January–February 2005, pp. 39–43.

Bruno, Greg, "The Role of the 'Sons of Iraq' in Improving Security," *Washington Post*, April 28, 2008. As of December 6, 2010: http://www.washingtonpost.com/wp-dyn/content/article/2008/04/28/AR2008042801120.html

Buchan, Glen, *Information War and the Air Force: Wave of the Future? Current Fad?* Santa Monica, Calif.: RAND Corporation, IP-149, 1996.

Chairman of the Joint Chiefs of Staff Instruction 3201.01, Joint Information Warfare Policy, January 2, 1996.

Chiarelli, Peter W., and Stephen M. Smith, "Learning from Our Modern Wars: The Imperatives of Preparing for a Dangerous Future," *Military Review*, September–October 2007, pp. 2–15.

Chief of Naval Personnel Public Affairs, "Information Dominance Corps Warfare Insignia Approved," press release, February 22, 2010. As of January 17, 2011: http://www.navy.mil/search/display.asp?story_id=51448

CJCSI—*see* Chairman of the Joint Chiefs of Staff Instruction.

Coran, Robert, *Boyd: The Fighter Pilot Who Changed the Art of War*, New York: Back Bay Books, May 10, 2004.

Csrnko, Thomas R., "Military Information Support Operations," memorandum to soldiers and civilians associated with the Psychological Operations Regiment, Ft. Bragg, N.C., June 23, 2010.

Cummings, Michael, and Eric Cummings, "5 Tips for Better Information Operations," *On Violence*, 2009. As of December 6, 2010: http://onviolence.com/?e=47

Democracy Now! "Twitter Crackdown: NYC Activist Arrested for Using Social Networking Site During G-20 Protest in Pittsburgh," October 6, 2009. As of December 6, 2010: http://www.democracynow.org/2009/10/6/twitter_crackdown_nyc_activist_arrested_for

Dempsey, GEN Martin, commanding general, U.S. Army Training and Doctrine Command, "Posturing the Army for Cyber, EW and IO as Dimensions of Full Spectrum Operations," memorandum to GEN Peter Chiarelli, U.S. Army Vice Chief of Staff, 2009.

DoD—*see* U.S. Department of Defense.

"DoD Launches POM-12 Study on Info Ops, Strategic Communication," *Inside the Pentagon*, May 20, 2010.

Dominique, Michael J., "Some Misconceptions Regarding Information Operations," *IO Journal*, May 2010.

Dorsett, VADM Jack, Deputy Chief of Naval Operations for Information Dominance, "Information Dominance and the U.S. Navy's Cyber Warfare Vision," briefing, 2010. As of January 17, 2011:
http://www.dtic.mil/ndia/2010SET/Dorsett.pdf

Elder, Lt Gen (ret.) Robert, review of draft of *Redefining Information Warfare Boundaries for an Army in a Wireless World*, October 13, 2010.

Electronic Frontier Foundation, "Man Arrested for Twittering Goes to Court, EFF Has the Documents," October 5, 2009. As of December 6, 2010:
http://www.eff.org/deeplinks/2009/10/
man-arrested-twittering-goes-court-eff-has-documen

Emery, Norman E., "Irregular Warfare Information Operations: Understanding the Role of People, Capabilities, and Effects," *Military Review*, November–December 2008, pp. 27–38.

Fitsanakis, Joseph, and Ian Allen, *Cell Wars: The Changing Landscape of Communications Intelligence*, Athens: Research Institute for European and America Studies, Research Paper No. 131, May 2009. As of May 19, 2009:
http://rieas.gr/images/RIEAS131.pdf

Fulghum, David A., "U.S. Navy Wants to Field Cyber-Attack System," *Aviation Week's DTI*, March 31, 2010a.

———, "Navy Pushes Cyber Options," *Aviation Week and Space Technology*, April 5, 2010, and April 29, 2010b.

———, "Advancing Stealth," *Aviation Week and Space Technology*, April 26, 2010c.

———, "Digital Demons," *Aviation Week and Space Technology*, Vol. 172, No. 18, May 10, 2010d.

———, "U.S. Army Pursues Afghanistan ISR Needs," *Aviation Week and Space Technology*, May 10, 2010e.

Gates, Robert M., Secretary of Defense, "Strategic Communication and Information Operations in the DoD," memorandum, January 25, 2011.

Gertz, Bill, "Inside the Ring," *Washington Times*, September 17, 2009.

Goldstein, Josh, *The Role of Digital Networked Technologies in the Ukrainian Orange Revolution*, Cambridge, Mass.: Berkman Center for Internet and Society, No. 2007-14, December 2007. As of December 6, 2010:
http://cyber.law.harvard.edu/sites/cyber.law.harvard.edu/files/Goldstein_
Ukraine_2007.pdf

Goodman, Amy, "Watch What You Tweet," *Truthdig*, October 6, 2009. As of December 8, 2010:
http://www.truthdig.com/report/item/20091006_watch_what_you_tweet/?ln

Gould, Joe, "Dempsey Delays Info Ops Manual Update to Settle Key Questions," *Inside Defense*, September 21, 2009.

Gray, David J., "Social Science and Social Policy: A Comment," *American Journal of Economics and Sociology*, 1989, pp. 307–309.

Greenmyer, John, "Say What You Mean . . .," *Dime Blog*, May 4, 2010. As of December 8, 2010:
http://www.carlisle.army.mil/dime/blog/archivedArticle.cfm?blog=dime&id=102

Headquarters, U.S. Department of the Army, *Information Operations*, Field Manual 100-6, Washington, D.C., August 1996.

———, *Strategic Departmental, and Operational IEW Operations*, Field Manual 34-37, Washington, D.C., draft, 1997a.

———, *Public Affairs Operations*, Field Manual 46-1, Washington, D.C., May 30, 1997b.

———, *Information Operations: Doctrine, Tactics, Techniques and Procedures*, Field Manual 3-13, Washington, D.C., November 2003.

———, *Psychological Operations*, Field Manual 3-05.30/Marine Corps Reference Publication 3-40.6, Washington, D.C., April 2005.

———, *Psychological Operations Tactics, Techniques, and Procedures*, Field Manual 3-05.301, Washington, D.C., August 30, 2007.

———, *Operations*, Field Manual 3-0, Washington, D.C., February 2008a.

———, *Knowledge Management Section*, Field Manual 6-01.1, Washington, D.C., August 2008b.

———, *Electronic Warfare in Operations*, Field Manual 3-36, Washington, D.C., February 25, 2009.

———, *Operations*, Field Manual 3-0, draft, Washington, D.C., April 30, 2010a.

———, *Army Electromagnetic Spectrum Operations*, Field Manual 6-02.70, Washington, D.C., May 20, 2010b.

Helmus, Todd C., Christopher Paul, and Russell W. Glenn, *Enlisting Madison Avenue: The Marketing Approach to Earning Popular Support in Theaters of Operation*, Santa Monica, Calif.: RAND Corporation, MG-607-JFCOM, 2007. As of December 8, 2010:
http://www.rand.org/pubs/monographs/MG607.html

Henderson, Eric, "Doctrine," *The Beacon*, Vol. 2 No. 1, January–February 2009.

Holley, I. B., "Of Saber Charges, Escort Fighters, and Spacecraft: The Search of Doctrine," *Air University Review*, September–October 1983.

HQDA—*see* Headquarters, U.S. Department of the Army.

Hura, Myron, review of draft of *Redefining Information Warfare Boundaries for an Army in a Wireless World*, December 2010.

"Iran's Twitter Revolution," *Washington Times*, June 16, 2009. As of December 6, 2010:
http://www.washingtontimes.com/news/2009/jun/16/irans-twitter-revolution

Jordan, Bryant, *Army Building Electronic Warfare Soldiers*, February 11, 2009. As of December 6, 2010:
http://defensetech.org/?s=Army+Building+Electronic+Warfare+Soldiers

Kelsall, Chris, "DON IT Workforce," briefing to integrated process team, September 2009.

Kruzel, John J., "Army Upgrades Its Electronic Warfare Training," American Forces Press Service, February 23, 2007. As of December 6, 2010:
http://www.army.mil/-news/2007/02/23/1947-army-upgrades-its-electronic-warfare-training/

Kuehl, Dan, "New Developments in IO," *IO Journal*, Vol. 1, No. 2, 3rd Quarter 2009, p. 6.

Lynn, William J. III, Deputy Secretary of Defense, "Responsible and Effective Use of Internet-Based Capabilities," Directive-Type Memorandum 09-026, incorporating change 1, September 16, 2010.

Morozov, Evgeny, "Moldova's Twitter Revolution," *Foreign Policy*, April 7, 2009. As of December 8, 2010:
http://neteffect.foreignpolicy.com/posts/2009/04/07/moldovas_twitter_revolution

Murphy, Dennis M., *Fighting Back: New Media and Military Operations*, Carlisle, Pa.: Center for Strategic Leadership, U.S. Army War College, November 2008.

———, *Talking the Talk: Why Warfighters Don't Understand Information Operations*, Carlisle, Pa.: Center for Strategic Leadership, U.S. Army War College, May 2009.

NATO—*see* North Atlantic Treaty Organization.

Nicholas, James, "Australia: Current Developments in Australian Army Information Operations," *IOSphere*, 2008, pp. 38–43.

North Atlantic Treaty Organization, "Allied Joint Doctrine for Information Operations," AJP 3–10, Brussels, 4th Study Draft, 1–3, 2007.

Paddock, Jr., Alfred, "PSYOP: On a Complete Change in Organization, Practice, and Doctrine," *Small Wars Journal*, June 26, 2010.

Paschall, Joseph F., "IO for Joe: Applying Strategic IO at the Tactical Level," *Field Artillery*, July–August 2005.

Paul, Christopher, *Information Operations: Doctrine and Practice: A Reference Handbook*, Westport, Conn.: Greenwood Publishing Group, 2008.

———, "'Strategic Communication' Is Vague: Say What You Mean," *Joint Force Quarterly*, No. 56, December 21, 2009a.

———, *Wither Strategic Communication? An Inventory of Current Proposals and Recommendations*, Santa Monica, Calif.: RAND Corporation, OP-250-RC, 2009b.

Pawlowski, Thomas J., ed., "Command, Control, Communications, Intelligence, Electronic Warfare Measures of Effectiveness (C3IEW MOE) Workshop," Ft. Leavenworth, Kan., October 20–23, 1992. As of December 6, 2010: http://www.dtic.mil/cgi-bin/GetTRDoc?AD=ADA331068&Location=U2&doc=GetTRDoc.pdf

Peterson, Robert, "Evolution of FM 34-40 to IEW Support to C2W," in *The Intel XXI Concept III: The Seven Intelligence Tasks*, 1997. As of December 8, 2010: http://www.fas.org/irp/agency/army/mipb/1997-1/cncp9701.htm

Porche, Isaac R. III, RAND Corporation, "Developing Army Capabilities for Cyber Operations," briefing to MG Rhett A Hernandez, 2007.

Porche, Isaac R. III, Jerry M. Sollinger, Bradley Wilson, Jeff Rothenberg, Joshua S. Caplan, and Joji Montelibano, *Bits on the Ground: Closing the Gaps in Defense of the Army's Networks*, Santa Monica, Calif.: RAND Corporation, 2010. Not available to the general public.

Rechtin, Eberhart, *System Architecting: Creating and Building Complex Systems*, Upper Saddle River, N.J.: Prentice Hall, 1991.

Richter, Walter E., "The Future of Information Operations," *Military Review*, January–February 2009, pp. 103–113.

Rohm, Fredric W., Jr., "Merging Information Operations and Psychological Operations," *Military Review*, January–February 2008, pp. 108–111.

Rosin, Randolph, "To Kill a Mockingbird: The Deconstruction of Information Operations," *Small Wars Journal*, August 17, 2009. As of December 6, 2010: http://smallwarsjournal.com/blog/2009/08/the-deconstruction-of-informat

Stack, Graham, "'Twitter Revolution' Moldovan Activist Goes into Hiding," *Guardian*, April 15, 2009. As of December 8, 2010: http://www.guardian.co.uk/world/2009/apr/15/moldova-activist-hiding-protests

Stern, Matthew, former commander, 1st IO Command, 2nd Battalion, personal communication with the authors, September 1, 2008.

TRAC—*see* U.S. Army Training and Doctrine Command Analysis Center.

TRADOC—*see* U.S. Army Training and Doctrine Command.

UK Ministry of Defence, Chiefs of Staff, *Information Operations*, Joint Warfare Publication 3–80, 2-1, Shrivenham, UK, 2002.

U.S. Army, "Technical Transitions: Roadmaps for IEW Systems," in *Army Science and Technology Master Plan (ASTMP 1997)*, Washington, D.C., March 21, 1997.

———, *U.S. Army Cyber/Electromagnetic Contest Capabilities-Based Assessment Study Plan*, V2.2, April 1, 2010.

U.S. Army Combined Arms Center, *Information and Cyberspace Symposium*, Ft. Leavenworth, Kan., April 15–18, 2008.

U.S. Army Information Proponent Office, "Director's Corner," *The Beacon*, Vol. 2, No. 2, March–April 2009a.

———, "Frequently Asked Questions," *The Beacon*, Vol. 2, No. 2, March–April 2009b.

U.S. Army Signal Center, *Electromagnetic Spectrum Operations Capability-Based Assessment*, 2010.

U.S. Army Training and Doctrine Command, *The United States Army Concept Capability Plan for Army Electronic Warfare Operations for the Future Modular Force 2015–2024*, Pamphlet 525-7-6, August 2007.

———, *The Army Capstone Concept Operational Adaptability—Operating Under Conditions of Uncertainty and Complexity in an Era of Persistent Conflict 2016–2028*, Pamphlet 525-3-0, December 21, 2009.

———, *The United States Army's Cyberspace Operations Concept Capability Plan 2016–2028*, Pamphlet 525-7-8, February 22, 2010a.

———, *United States Army Operating Concept 2016–2028*, Pamphlet 525-3-1, August 19, 2010b.

U.S. Army Training and Doctrine Command Analysis Center, *Electronic Warfare Capability-Based Assessment*, final report, Ft. Leavenworth, Kan., TRAC-F-TR-09-011, 2009.

U.S. Army Training and Doctrine Command Combined Arms Center, Capability Development Integration Directorate, "Army Cyber/Electromagnetic Contest Capabilities-Based Assessment (C/EM CBA) Functional Solutions Analysis (FSA) Workshop #1," 2009.

———, *The Cyber/Electromagnetic Contest, A White Paper*, Ft. Leavenworth, Kan., 2010.

U.S. Department of Defense, *Information Operations Roadmap*, Washington, D.C., October 30, 2003.

U.S. Department of Defense Directive 8500.01E, Information Assurance, October 24, 2007.

U.S. Department of Defense Instruction 8500.2, Information Assurance (IA) Implementation, February 6, 2003.

U.S. Joint Chiefs of Staff, *C4I for the Warrior: A 1995 Progress Report*, Washington, D.C., 1995.

———, *Joint Doctrine for Information Operations*, Joint Publication 3-13, Washington, D.C., October 9, 1998.

———, *Information Operations*, Joint Publication 3-13, Washington, D.C., February 13, 2006a.

———, *The National Military Strategy for Cyberspace Operations*, Washington, D.C., December 2006b.

———, *Electronic Warfare*, Joint Publication 3-13.1, Washington, D.C., January 25, 2007.

———, *Public Affairs*, Joint Publication 3-61, Washington, D.C., August 25, 2010a.

———, *Department of Defense Dictionary of Military and Associated Terms*, Joint Publication 1-02, Washington, D.C., April 12, 2001, as amended through September 30, 2010b.

U.S. Navy, "Information Warfare," information warfare officer career description, web page, undated. As of December 8, 2010:
http://www.navy.com/navy/careers/information-and-technology/information-warfare.html

Vanden Brook, Tom, "Signals Foil IEDs, but Also Radios," *USA Today*, January 22, 2007. As of December 6, 2010:
http://www.usatoday.com/news/washington/2007-01-22-jamming_x.htm

Wass de Czege, Huba, "Rethinking 'IO': Complex Operations in the Information Age," *Small Wars Journal*, July 4, 2008.

Wegner, Neal, *The Intel XXI Concept III: The Seven Intelligence Tasks, Military Intelligence Professional Bulletin*, April–June 1996.

Whisenhunt, John, ed., "The View from the Army IO Proponent: Colonel David Haught Interview," *IO Sphere*, Winter 2009, pp. 28–31.

Williams, Michael L., and Marc J. Romanych, "Information Operations: Not Just a Set of Capabilities," *IO Journal*, December 2009, pp. 18–22.